BIOLOGY
• PRINCIPLES & EXPLORATIONS •

Student Review Guide

HOLT, RINEHART AND WINSTON

A Harcourt Classroom Education Company

Austin • New York • Orlando • Atlanta • San Francisco • Boston • Dallas • Toronto • London

To the Student

This Student Review Guide is designed to help you succeed in your work with *Biology: Principles and Explorations*. You will find various activities that you can work on independently as you study and review the textbook. In this book, you will find three different kinds of worksheets:

Test Preparation Pretests
The Test Preparation Pretests help you learn and remember the contents of the textbook. The Pretests also help you evaluate whether you can recall the facts presented in each chapter. The Pretests contain multiple choice, matching, completion, true-false, and short answer exercises.

Vocabulary Worksheets
The Vocabulary Worksheets consist of crossword puzzles, word scrambles, matching exercises, and other activities designed to help you become familiar with and understand the key terms from each chapter of the textbook.

Science Skills Worksheets (Chapters 1–20, 24, 28, 33, and 38)
The Science Skills Worksheets focus on skills—such as analyzing and interpreting data, using maps, analyzing experiments, and interpreting graphics—that enable you to read and think scientifically. These worksheets also help you understand the core content of the textbook. One worksheet is provided for each of the 19 chapters of Part 1: Principles (Units 1–4) of the textbook. In addition, one worksheet is provided for each of the five introductory chapters of Part 2: Explorations (Units 5–9) of the textbook.

Illustration credits: 11, Kristy Sprott; 12, Kristy Sprott; 41, Rosiland Solomon; 46, Rosiland Solomon; 49, Rosiland Solomon; 52, Rosiland Solomon; 92, Kristy Sprott; 97, Pedro Julio Gonzalez/Melissa Turk & The Artist Network; 173(tl), Joel Floyd; 173(tr), Rosiland Solomon; 235, David Kelley; 239, Pedro Julio Gonzalez/Melissa Turk & The Artist Network; 254, Morgan-Cain & Associates.

Printed in the United States of America

ISBN 0-03-054366-5

7 8 9   18   05 04 03 02

Table of Contents

CHAPTER

1 TEST PREP PRETEST

Biology and You

In the space provided, write the letter of the term or phrase that best completes each statement or best answers each question.

_____ 1. When David Bradford visited the mountain lakes of Sequoia-Kings Canyon National Parks in the summer of 1988, he discovered that the population of the *Rana muscosa* had

    a. increased slightly.   **c.** stayed the same.
    b. increased dramatically.   **d.** decreased dramatically.

_____ 2. Control and experimental groups are identical except for the

    a. dependent variable.   **c.** independent variable.
    b. group size.   **d.** conclusions.

_____ 3. The world's tropical forests should be saved because

    a. the human population has passed 6 billion people.
    b. 1,000 plant and animal species live in them.
    c. they contain food and medicine plants.
    d. None of the above

_____ 4. Which of the following steps in a scientific investigation is usually taken first?

    a. experimenting   **c.** theorizing
    b. hypothesizing   **d.** observing

_____ 5. In a scientific investigation, a possible explanation is called a(n)

    a. hypothesis.   **c.** observation.
    b. inference.   **d.** analysis.

_____ 6. Comparison of data gathered under conditions in which the key factor is not allowed to change is part of the process of

    a. hypothesizing.   **c.** observing.
    b. predicting.   **d.** analyzing.

_____ 7. A collection of related hypotheses that have been tested many times is called a(n)

    a. prediction.   **c.** theory.
    b. observation.   **d.** insight.

_____ 8. The study of life is called

    a. ecology.   **c.** morphology.
    b. biology.   **d.** phylogeny.

_____ 9. What properties do all living things exhibit?

    a. cellular organization, metabolism, homeostasis, reproduction, and heredity
    b. multicellular organization, metabolism, homeostasis, reproduction, and heredity
    c. photosynthesis, metabolism, homeostasis, reproduction, and heredity
    d. cellular organization, photosynthesis, homeostasis, reproduction, and heredity

_____ 10. Which of the following is NOT a major environmental concern caused by the growing human population?
  a. pollution of the atmosphere   **c.** waste disposal
  b. extinction of species   **d.** reduction of oil reserves

_____ 11. The demand for food to feed the world's population is going to
  a. decrease.
  b. increase.
  c. stay the same.
  d. decrease until the year 2010, then increase.

_____ 12. Almost all lung cancers can be prevented by
  a. changing your diet.
  b. eliminating the use of tobacco.
  c. staying out of the sun.
  d. exercising regularly.

• • • • • • • • • • • • • •

Circle T *if the statement is true or* F *if it is false.*

T   F   **13.** Amphibians are very sensitive to changes in their environment because their moist skin can absorb chemicals from pond water.

T   F   **14.** In John Harte's experiment, the pH level was the dependent variable because it affected the number of salamanders that hatched.

T   F   **15.** John Harte concluded that a steady decrease in the tiger salamander population during the 1980s was caused by the thinning of the protective ozone layer.

T   F   **16.** The stages of a scientific investigation are collecting observations, asking questions, forming hypotheses and making predictions, confirming predictions, and drawing conclusions.

T   F   **17.** A statement of what you expect to happen if a hypothesis is correct is called a prediction.

T   F   **18.** The control is the part of the experiment in which the key factor is changed.

T   F   **19.** The study of interactions of living organisms with one another and their environment is called ecology.

T   F   **20.** Only living things exhibit the properties of movement and complexity.

T   F   **21.** Living things maintain relatively stable internal conditions through a process called homeostasis.

T   F   **22.** Only multicellular organisms exhibit cellular organization.

T   F   **23.** Genetic engineers are transplanting genes from one crop plant to another to create crop plants that are resistant to insects.

T   F   **24.** All mutations are passed on to other generations.

• • • • • • • • • • • • • • •

Complete each statement by writing the correct term or phrase in the space provided.

25. John Harte determined that snowmelt in the mountains of west-central Colorado

released high levels of _____ .

26. A scientist collects data to test a(n) _____ .

27. HIV is a virus that destroys the _____ _____ .

28. A single variable is held constant in a(n) _____

_____ .

29. All living things pass on genetic information from parent to offspring through a

process known as _____ .

30. _____ is the change in a species' inherited traits over time.

31. _____ is a disorder of cells in which the normal controls on
growth have been damaged and the cells divide unchecked within the body.

32. _____ _____ is an often fatal genetic
disorder in which thick mucus builds up in the lungs and other organs.

33. The basic unit of all organisms that can carry out all life processes is the

_____ .

34. An organism with many of these units is referred to as _____ .

• • • • • • • • • • • • • • •
Read each question, and write your answer in the space provided.

35. Define the first stage of a scientific investigation—observation.

_____

_____

36. How will biological studies of tropical forests help save the forests?

_____

_____

_____

37. Describe interdependence in a biological community.

_____

_____

_____

38. How does the publication of research in a scientific journal benefit scientists?

_____

_____

_____

_____

39. Explain why there is no absolute certainty or scientific "truth" in a theory.

_____

_____

_____

40. What is the role of metabolism in life?

_____

_____

_____

_____

_____

41. Why is reproduction an essential part of living?

_____

_____

_____

42. Harte's research supported his hypothesis that acid in melting snow killed salamanders. Why is this hypothesis not yet accepted as a theory?

_____

_____

_____

_____

43. Briefly describe the efforts to increase food production.

_____

_____

_____

_____

44. Briefly describe recent scientific advances in finding a cure for cystic fibrosis.

_____

_____

_____

_____

Name_____ Date _____ Class _____

CHAPTER
1 VOCABULARY

Biology and You

In the space provided, explain how the terms in each pair differ in meaning.

1. cell, metabolism

_____

_____

2. heredity, mutation

_____

_____

3. natural selection, evolution

_____

_____

4. hypothesis, prediction

_____

_____

5. independent variable, dependent variable

_____

_____

6. theory, observation

_____

_____

7. cystic fibrosis, gene

_____

_____

(continued on next page)

In the space provided, write the letter of the description that best matches the term or phrase.

_____ 8. biology

_____ 9. control group

_____ 10. homeostasis

_____ 11. species

_____ 12. reproduction

_____ 13. ecology

_____ 14. experiment

_____ 15. HIV

_____ 16. cancer

_____ 17. pH

a. a group of organisms that are genetically similar and can produce fertile offspring

b. state of constant internal conditions

c. the study of living things

d. receives no experimental treatment

e. organisms make more of their own kind

f. the study of the interactions of living organisms

g. planned procedure to test a hypothesis

h. relative measure of the hydrogen ion concentration within a solution

i. virus that causes AIDS

j. disorder caused by uncontrolled cell division

CHAPTER

1 **SCIENCE SKILLS: SEQUENCING/RELATING INFORMATION**

Biology and You

Scientists at a major university became concerned by recent reports of severe insomnia in a number of people with previously normal sleep patterns. These scientists decided to undertake a scientific investigation to try to determine the cause of this behavior. This scientific investigation had six stages. The paragraphs below, labeled *A–F*, describe the stages of this investigation, but the sequence of the stages is incorrect.

Sequence the stages properly by listing the letters of the paragraphs below in the correct order in the chart on page 8. Then decide which step in the scientific process each paragraph describes. In the space provided in the chart, write the step next to the number.

A. The scientists gathered 35 volunteers who agreed to eat 3 meals a day while living at the university research center for 30 days. These volunteers had no previous history of insomnia. They were aware that the meals would be made up of some foods containing the fat substitute, and they had been informed of the potential risks involved. Within 14 days, 17 of the volunteers were having trouble sleeping at night, and within 21 days, the entire group was showing signs of insomnia.

B. The scientists stated that if the chemical in the fat substitute was reacting with chemicals in the brain to cause the insomnia in the subjects being investigated, it should have a similar effect on other people.

C. The scientists gathered 80 volunteers who agreed to eat three meals a day while living at the university research center for 60 days. These volunteers had no previous history of insomnia. They were aware that the meals would be made up of some foods containing the fat substitute, and they had been informed of the potential risks involved. The scientists divided the volunteers into two groups. Group A consisted of 40 volunteers who were fed meals with foods containing the fat substitute. Group B consisted of 40 volunteers who were fed meals that did not contain the fat substitute. At the end of 60 days, the 40 volunteers in Group A were suffering from insomnia; the 40 volunteers in Group B were sleeping normally. Moreover, within 7 days of discontinuing the diet containing the fat substitute, the sleep patterns of the members of Group A returned to normal.

D. Because the fat substitute contained a chemical not typically found in the human diet, the scientists thought that this chemical caused the insomnia by reacting with other chemicals in the brain.

E. The scientists agreed that the chemical in the fat substitute was causing insomnia in people who ate foods containing this additive.

F. The scientists began to track the personal habits of 127 people who had recently reported the onset of insomnia, hoping to find a clue to the cause of this sleep disorder. Information about previous and current sleep patterns, exercise routines, stress at home and work, eating habits, and other criteria was gathered over a 6-month period. The only common element in the lives of all of the subjects was the consumption of foods containing a new fat substitute.

Correct order	Step in the scientific process
1. _____	_____
2. _____	_____
3. _____	_____
4. _____	_____
5. _____	_____
6. _____	_____

CHAPTER

2 **TEST PREP PRETEST**

Chemistry of Life

In the space provided, write the letter of the term or phrase that
best completes each statement or best answers each question.

_____ 1. An element is defined as a pure substance because it is made of only one kind of

    a. compound.   **c.** atom.
    b. molecule.   **d.** ion.

_____ 2. In an atom with the same number of electrons and protons, the electrical charge is

    a. neutral.   **c.** negative.
    b. positive.   **d.** impossible to determine.

_____ 3. Atoms bond with other atoms because

    a. neutrons are attracted to other neutrons.
    b. bonding helps atoms become more stable.
    c. the nuclei of atoms are attracted to each other.
    d. bonding makes an element more pure.

_____ 4. When a group of atoms is held together by covalent bonds,

    a. a molecule is formed.   **c.** an element is formed.
    b. an ion is formed.   **d.** a polar compound is
                               formed.

_____ 5. Ionic bonds occur between

    a. a neutral ion and a negatively charged ion.
    b. two nonpolar molecules.
    c. two negatively charged ions.
    d. a negatively charged ion and a positively charged ion.

_____ 6. Acids and bases differ in that

    a. bases dissolved in water form more hydrogen ions than do acids dissolved in water.
    b. acids dissolved in water form more hydrogen ions than do bases dissolved in water.
    c. acids dissolved in water form more hydroxide ions than do bases dissolved in water.
    d. bases have a lower pH than do acids.

_____ 7. When food molecules are broken down inside cells, some of the energy in the molecules is stored temporarily in

    a. proteins.   **c.** DNA.
    b. RNA.   **d.** ATP.

_____ 8. Which of the following groups of terms is associated with carbohydrates?

    a. monosaccharide, glycogen, cellulose
    b. monosaccharide, cellulose, lipid
    c. disaccharide, polysaccharide, steroid
    d. polysaccharide, amino acid, collagen

_____ 9. The speed of a chemical reaction is increased by
   a. an enzyme.   **c.** glucose.
   b. the reactant.   **d.** ATP.

_____ 10. An enzyme's active site allows the enzyme to attach to
   a. a molecule of water.   **c.** another enzyme.
   b. a cell.   **d.** a substrate.

• • • • • • • • • • • • • • • •

In the space provided, write the letter of the description that best matches the term or phrase.

_____ 11. atom

_____ 12. element

_____ 13. compound

_____ 14. molecule

_____ 15. electron

_____ 16. hydrogen bond

_____ 17. cohesion

_____ 18. adhesion

_____ 19. acid

_____ 20. pH

_____ 21. carbohydrates

_____ 22. lipids

_____ 23. saturated fatty acid

_____ 24. protein

_____ 25. nucleic acid

_____ 26. catalyst

a. an attraction between substances of the same kind

b. a class of organic compounds made of carbon, hydrogen, and oxygen atoms in a 1:2:1 ratio

c. the smallest unit of matter that cannot be broken down by chemical means

d. a measure of the concentration of hydrogen ions in solutions

e. solid at room temperature

f. a substance that decreases the activation energy of a chemical reaction

g. a class of organic compounds that includes fats, steroids, and waxes

h. an attraction between different substances

i. a substance made of only one kind of atom

j. a weak attraction between polar molecules

k. a molecular chain of amino acids

l. a negatively charged atomic particle

m. a substance made when atoms of two or more different elements join together

n. a group of atoms held together by covalent bonds

o. a compound that forms hydrogen ions when dissolved in water

p. a molecular chain of nucleotides

• • • • • • • • • • • • • • • •

Complete each statement by writing the correct term or phrase in the space provided.

27. An atom with more electrons than protons would have a(n)

   _____ charge.

28. The weak chemical attraction between water molecules is a(n)

   _____ _____ , while the stronger chemical

   bond between the atoms of each water molecule is a(n) _____

   _____ .

29. An atom or molecule that has gained or lost one or more electrons is called a(n)

_____ .

30. On the pH scale, vinegar is a(n) _____ and ammonia is a(n)

_____ .

31. Two _____ that store energy are starch—produced by plants, and glycogen—produced by animals.

32. _____ , one of the major classes of carbon compounds, are

nonpolar and include _____ , _____ ,

steroids, and _____ .

33. A(n) _____ consists of a sugar, a base, and a phosphate group.

34. _____ _____ , such as DNA and RNA, are
made of long chains of nucleotides.

35. When sodium chloride is dissolved in water, a chemical reaction occurs in which

sodium chloride is the _____ and sodium ions and chloride ions

are the _____ .

36. Enzymes are a type of _____ , which reduces the activation
energy of a chemical reaction.

37. A substrate attaches to the _____ _____ of
an enzyme.

38. _____ and _____ can affect enzyme activity.

• • • • • • • • • • • • •
Read each question, and write your answer in the space provided.

Questions 39 and 40 refer to the figure at right.

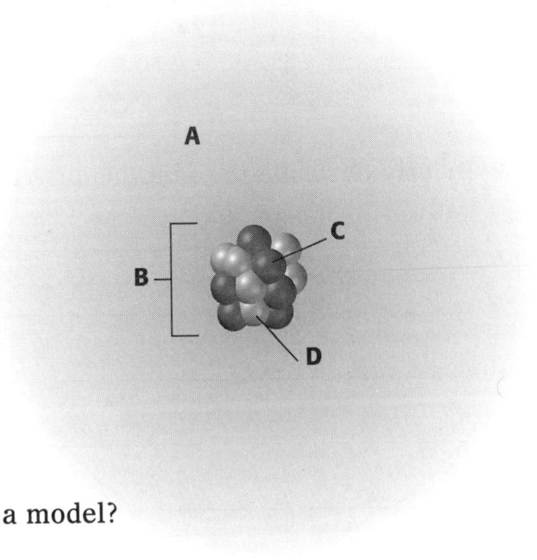

39. What does this figure represent? Identify the structures labeled *A*, *B*, *C*, and *D*.

_____

_____

_____

40. Why is the figure at right considered to be only a model?

_____

_____

_____

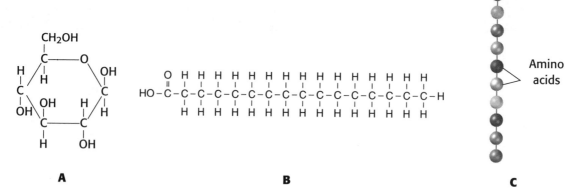

A   **B**   **C**

41. Identify the principal class of organic compound represented by each of the molecules shown above.

_____

_____

_____

42. Identify each of the molecules shown above.

_____

_____

_____

43. Where is each of these molecules found? Give an example for each one.

_____

_____

_____

_____

44. Briefly describe the function of ATP in cells.

_____

_____

_____

_____

_____

CHAPTER

2 **VOCABULARY**

Chemistry of Life

In the space provided, write the letter of the description that best matches the term or phrase.

_____ 1. ion

_____ 2. atom

_____ 3. compound

_____ 4. molecule

_____ 5. covalent bond

_____ 6. ionic bond

_____ 7. element

_____ 8. solution

a. smallest unit of matter that cannot be broken down by chemical means

b. a substance made of the joined atoms of two or more different elements

c. atom or molecule that has lost or gained one or more electrons

d. a substance made of only one type of atom

e. one substance evenly distributed in another

f. chemical bond in which electrons are shared

g. a group of atoms held together by covalent bonds

h. attraction between oppositely charged ions

Complete each statement by writing the correct term or phrase in the space provided.

9. A(n) _____ is a substance on which an enzyme acts during a chemical reaction.

10. An organic compound with a ratio of one carbon atom to two hydrogen atoms to one

    oxygen atom is a(n) _____ .

11. Glucose is a(n) _____ that is a major source of energy in cells.

12. A(n) _____ is an organic compound that is not soluble in water.

13. A(n) _____ is a long chain of amino acids.

14. Subunits of DNA and RNA are called _____ .

15. DNA is a(n) _____ _____ that encodes protein sequences.

(continued on next page)

In the space provided, explain how the terms in each pair differ in meaning.

16. acid, base

_____

_____

17. cohesion, adhesion

_____

_____

_____

18. enzyme, active site

_____

_____

_____

19. energy, activation energy

_____

_____

20. DNA, RNA

_____

_____

_____

21. ATP, carbohydrate

_____

_____

CHAPTER

2 **SCIENCE SKILLS:** ANALYZING INFORMATION/INTERPRETING GRAPHS

Chemistry of Life

Dirt sticks to the body either by becoming trapped in microscopic wrinkles in the skin or, if the dirt is moist, by adhering to the body. Sometimes the natural oils on skin will give the dirt an oily coating. In such cases, water alone will not remove the dirt, but soap and water will. Use the information below and your understanding of polarity and chemical bonding to answer questions 1–3.

A. A soap molecule is long with one end attracted to oil molecules.

B. One end of a soap molecule is polar and the other end is nonpolar.

C. Soap will dissolve, and the soap molecules will float freely in water.

D. A sewing needle will rest upon the surface of water. If powdered laundry detergent is gently sprinkled near the needle, the needle will eventually sink.

Read each question, and write your answer in the space provided.

1. Explain why adding soap to water will help remove dirt and oil.

_____

_____

_____

_____

_____

2. Why does the needle float on the water?

_____

_____

_____

3. Why does the needle sink after soap is added to the water?

_____

_____

_____

_____

The graph below shows the rate of enzyme activity in relation to pH for two enzymes, pepsin and trypsin. Both enzymes break down molecules in food taken into the human body, but the enzymes act in series. Pepsin breaks some bonds in very large molecules. Trypsin acts on the fragments produced by the action of pepsin, breaking them into even smaller units. Use the graph to answer questions 4–8 below.

Enzymes and pH

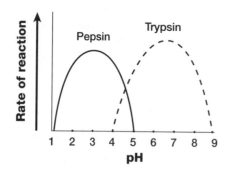

Read each question, and write your answer in the space provided.

4. The liquid in the stomach has a pH of about 2. Which of the two enzymes would be active in the stomach?

_____

_____

5. The liquid in the small intestine has a pH of about 8. Which of the two enzymes would be active in the small intestine?

_____

_____

6. What must happen to the liquid as it passes from the stomach to the small intestine for digestion to occur normally?

_____

_____

_____

7. Consider the data on the relationship between pH and enzyme activity shown in the graph. Do enzymes typically function only at a specific pH, or can they function at a range of pH values?

_____

_____

_____

8. Can pepsin and trypsin function in the same environment? Explain.

_____

_____

CHAPTER
3 **TEST PREP PRETEST**

Cell Structure

In the space provided, write the letter of the term or phrase that best completes each statement or best answers each question.

_____ 1. Which scientist determined that cells come only from other cells?
      a. van Leeuwenhoek   **c.** Schwann
      b. Schleiden   **d.** Virchow

_____ 2. The surface-area-to-volume ratio of a small cell is
      a. greater than that of a larger cell.
      b. less than that of a larger cell.
      c. equal to that of a larger cell.
      d. not affected by the cell's size.

_____ 3. In the metric system, a micrometer (μm) is equal to 0.000001 (one millionth) of a
      a. kilometer (km).   **c.** centimeter (cm).
      b. meter (m).   **d.** millimeter (mm).

_____ 4. A protein on the cell membrane that recognizes and binds to specific substances is called a(n)
      a. marker protein.   **c.** enzyme.
      b. receptor protein.   **d.** transport protein.

_____ 5. As a cell becomes smaller, its surface-area-to-volume ratio
      a. increases.   **c.** stays the same.
      b. decreases.   **d.** becomes less important.

_____ 6. Which of the following instruments produces highly magnified three-dimensional images of a cell's surface?
      a. hand lens   **c.** SEM
      b. light microscope   **d.** TEM

_____ 7. In a eukaryotic cell, mitochondria
      a. transport materials.   **c.** produce ATP.
      b. make proteins.   **d.** control cell division.

Questions 8 and 9 refer to the figure at right.

_____ 8. The cell in the figure at right is a(n)
      a. prokaryotic cell.
      b. eukaryotic cell.
      c. plant cell.
      d. Both (b) and (c)

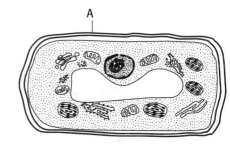

_____ 9. The structure labeled *A*
      a. supports the cell.
      b. protects the cell.
      c. surrounds the cell membrane.
      d. All of the above

In the space provided, write the letter of the description that
best matches the term or phrase.

_____ 10. cell membrane

_____ 11. central vacuole

_____ 12. chloroplasts

_____ 13. cytoplasm

_____ 14. electron microscope

_____ 15. Golgi apparatus

_____ 16. light microscope

_____ 17. lipid bilayer

_____ 18. lysosomes

_____ 19. magnification

_____ 20. mitochondrion

_____ 21. nucleus

_____ 22. organelles

_____ 23. phospholipid

_____ 24. resolution

_____ 25. ribosomes

_____ 26. scanning tunneling microscope

_____ 27. SI

_____ 28. vesicle

a. making an image look larger than its actual size

b. a boundary that encloses the cell, separating it from its surroundings

c. cellular structures responsible for protein production

d. double layer of phospholipids

e. interior of a cell

f. uses a beam of electrons to form an image of a specimen

g. houses the cell's DNA

h. in plant cells, a large, membrane-bound sac that stores water, nutrients, or other substances

i. a system of measurement based on powers of 10

j. light passes through one or more lenses to produce a magnified image of a specimen

k. a lipid made of a phosphate group and two fatty acids

l. a measure of an image's clarity

m. an organelle that produces ATP

n. in plant cells, organelles that use light to make organic compounds

o. the cell's packaging and distribution center

p. small organelles that contain the cell's digestive enzymes

q. a small, membrane-bound sac

r. specialized structures within a cell

s. produces a computer-generated three-dimensional image of an object's surface

Complete each statement by writing the correct term or phrase
in the space provided.

29. A cell's boundary is called the _____ _____ .

30. In a bacterium, the _____ _____ provides structure and support.

31. _____ _____ in the cell membrane aid the movement of substances into and out of a cell.

32. In plant cells, rigidity is provided by a large, membrane-bound sac called the

_____ _____ .

33. _____ _____ allow certain substances to pass into and out of the nucleus of a cell.

34. _____ are vesicles that contain a cell's digestive enzymes.

35. The "head" of a phospholipid is _____ , so it is attracted to

water, while the "tails" are _____ , so they are repelled by water.

36. The cytoskeleton is a network of protein fibers that support the shape of

a cell and may be involved in the movement of _____ .

37. If a compound microscope has a 50× objective lens and a 10× ocular lens,

a viewed image appears _____ times larger than its actual size.

38. Mitochondria contain their own _____ , so they can produce their own proteins.

Questions 39–45 refer to the figure below.

39. The structure labeled *A* is the _____ _____ .

40. The organelle labeled *B* is the _____ _____ .

41. The structure labeled *C* is the _____ _____ .

42. The structure labeled *D* is the _____ _____ .

43. The organelle labeled *E* is the _____ _____ .

44. The organelle labeled *F* is a(n) _____ .

45. The organelle labeled *G* is a(n) _____ .

Read each question, and write your answer in the space provided.

46. Explain the role of DNA in cells.

_____

_____

_____

_____

47. Explain how magnification and resolution are related.

_____

_____

_____

_____

_____

48. List the three parts of the cell theory.

_____

_____

_____

49. List the primary differences between prokaryotic cells and eukaryotic cells.

_____

_____

_____

_____

_____

50. Distinguish between plant cells and animal cells.

_____

_____

_____

_____

_____

Name_____ Date _____ Class _____

Cell Structure

In the blanks provided, fill in the letters of the term or phrase being described.

1. uses light to produce a   _ _ _ _ __T__ __M__ _ _ _ _ _ _ _
   magnified image

2. uses electrons to form a   _ _ _ __C__ _ _ _ _ __M__ _ _ _ _ _ _ _ _
   magnified image

3. when an image appears larger   _ _ __G__ _ _ _ _ _ _ _ _

4. measure of clarity of image   __R__ _ _ _ _ _ _ _ _ _

5. produces three-dimensional   _ _ _ __N__ _ _ _ _ _ __N__ _ _ _ _
   images of living organisms   _ __I__ _ _ _ _

6. all living things are made   _ _ _ __L__ _ __H__ _ _ _ _
   of cells

7. regulates what enters and   _ _ _ _ __L__ _ _ __M__ _ _ _ _ _ _
   leaves a cell

8. structure on which proteins   _ _ __B__ _ _ _ _ _
   are made

9. single-celled organism that   _ _ _ __K__ _ _ _ _ _ _
   lacks a nucleus

10. protrude from cell's surface   __F__ _ _ _ _ _ _ _
    and enable movement

11. organism whose cells each   _ __U__ _ _ _ _ _ _ _
    have a nucleus

12. houses the cell's DNA   _ _ __C__ _ _ _ _

13. carries out specific activities   _ _ __G__ _ _ _ _ _ _

14. hairlike structures   __C__ _ _ _ _

15. interior of cell   _ _ _ _ _ __P__ _ _ _

16. keeps cell membrane from   _ _ _ _ __S__ _ _ _ _ _ _
    collapsing

17. has a polar "head" and   _ _ _ __S__ _ _ _ _ _ _ _
    nonpolar "tails"

18. double layer of phospholipids   _ _ _ _ __D__ _ _ __L__ _ _ _ _

(continued on next page)

Complete each statement by writing the correct term or phrase
in the space provided.

19. The _____ _____ is an extensive system of internal membranes that moves proteins and other substances through the cell.

20. A(n) _____ is a small, membrane-bound sac.

21. The _____ _____ is the packaging and distribution center of the cell.

22. _____ are the organelles that contain the cell's digestive enzymes.

23. _____ transfer energy from organic compounds to ATP.

24. _____ are organelles that use light energy to make carbohydrates from carbon dioxide and water.

25. The _____ _____ stores water and may contain many substances, such as ions, nutrients, and wastes.

26. The cell membrane of a plant is surrounded by a thick _____

_____ , which supports and protects the cell.

Name_____ Date _____ Class _____

Cell Structure

Biology students were working on a class project. They prepared copies of transmission electron micrographs of a bacterium, a plant cell, and an animal cell for display in their classroom. Unfortunately, the pictures were not labeled and got mixed up. Help these students correctly identify the cells and cell structures. Use the figures below to answer questions 1–4.

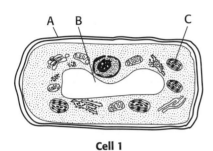

Cell 1

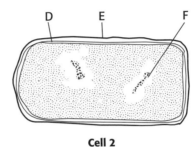

Cell 2

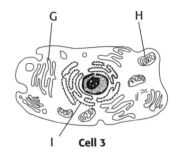

Cell 3

Read each question, and write your answer in the space provided.

1. In the space provided, write the names of each cell's labeled structures *(A–I)*. Using this information, write the identity of each cell—bacterium, plant cell, or animal cell—in the space provided.

   Cell 1 identity_____

   A. _____

   B. _____

   C. _____

   Cell 2 identity_____

   D. _____

   E. _____

   F. _____

   Cell 3 identity_____

   G. _____

   H. _____

   I. _____

2. Are these cells prokaryotic or eukaryotic?

_____

_____

_____

3. What are the primary differences between the three cells? What characteristics do they share?

_____

_____

_____

_____

_____

_____

_____

_____

_____

_____

4. The mitochondria in eukaryotic cells closely resemble prokaryotic organisms. In addition, mitochondria contain their own distinctive DNA. Develop a hypothesis for the ancestry of eukaryotic cells based on this information and your work above.

_____

_____

_____

_____

_____

_____

_____

Name_____ Date _____ Class _____

Cells and Their Environment

In the space provided, write the letter of the term or phrase that best completes each statement or best answers each question.

_____ 1. When a receptor protein in a cell membrane acts as an enzyme, the receptor protein
   a. changes its shape to allow the signal molecule to enter the cell.
   b. causes chemical changes in the cell.
   c. activates a second messenger that acts as a signal molecule within the cell.
   d. changes the permeability of the cell membrane.

_____ 2. Which of the following is NOT a characteristic of an ion channel?
   a. It extends from one side of the cell membrane to the other.
   b. It may or may not have a gate.
   c. It is polar, so charged substances, such as ions, can pass through the nonpolar lipid bilayer.
   d. It allows ions to move against their concentration gradient.

_____ 3. In the cell membrane, ion channels serve as
   a. food molecules.   **c.** information receivers.
   b. cell identifiers.   **d.** passageways.

_____ 4. The diffusion of water through a selectively permeable membrane is called
   a. exocytosis.   **c.** active transport.
   b. osmosis.   **d.** endocytosis.

_____ 5. If the "free" water molecule concentration outside a cell is higher than that inside the cell, the solution outside of the cell is
   a. isotonic.
   b. hypertonic.
   c. hypotonic.
   d. None of the above

_____ 6. Which of the following is an example of osmosis?
   a. the movement of ions from an area of high concentration to an area of lower concentration
   b. the movement of ions from an area of low concentration to an area of higher concentration
   c. the movement of "free" water molecules from an area of high concentration to an area of lower concentration
   d. the movement of "free" water molecules from an area of low concentration to an area of higher concentration

_____ 7. When particles move out of a cell through facilitated diffusion, the cell
   a. gains energy.
   b. uses energy.
   c. first gains and then uses energy.
   d. does not use energy.

_____ 8. When a cell uses energy to transport a particle through the cell membrane to an area of higher concentration, the cell is using

    a. diffusion.   **c.** osmosis.

    b. active transport.   **d.** facilitated diffusion.

_____ 9. Which of the following is an example of active transport?

    a. equilibrium   **c.** facilitated diffusion

    b. sodium-potassium pump   **d.** osmosis

_____ 10. The excretion of materials to the outside of a cell by discharging them from vesicles is called

    a. exocytosis.   **c.** osmosis.

    b. endocytosis.   **d.** diffusion.

_____ 11. If someone spills perfume in one room, people can soon smell it in an adjacent room. This is an example of

    a. facilitated diffusion.   **c.** diffusion.

    b. osmosis.   **d.** active transport.

_____ 12. Which of the following CANNOT occur when a signal molecule binds to a receptor protein on a cell's surface?

    a. The receptor can open an ion channel in the cell membrane.

    b. The receptor can act as an enzyme, causing chemical changes in the cytoplasm.

    c. The receptor can cause the formation of a second messenger.

    d. The receptor can transport the signal molecule into the cell through endocytosis.

_____ 13. The mechanism that prevents sodium ions from building up inside the cell is called

    a. the sodium-potassium pump.   **c.** diffusion.

    b. endocytosis.   **d.** exocytosis.

• • • • • • • • • • • • • • •

Circle T _if the statement is true or_ F _if it is false._

T   F   **14.** The cell membrane includes a triple layer of phospholipids.

T   F   **15.** Osmosis and diffusion are both examples of passive transport.

T   F   **16.** The membrane of a cell is selectively permeable.

T   F   **17.** The sodium-potassium pump requires no energy to function.

T   F   **18.** Diffusion requires energy from the cell.

T   F   **19.** A gated ion channel in the cell membrane may be opened or closed.

Question 20 refers to the figure at right.

T   F   **20.** The process shown in the figure at right is exocytosis.

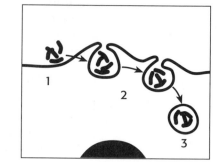

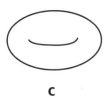

A   B   C

Complete each statement by writing the correct term or phrase in the space provided.

21. Figure A illustrates a cell in a(n) _____ solution.

22. Figure B illustrates a cell in a(n) _____ solution.

23. Figure C illustrates a cell in a(n) _____ solution.

24. When a substance moves from an area of low concentration to an area of higher concentration, the substance moves _____ its concentration gradient.

25. Cell-surface proteins allow a cell to _____ with other cells.

26. A(n) _____ protein binds to a specific signal molecule.

27. When the concentration of dissolved particles is the same throughout a solution, the system is said to be in _____ .

28. _____ _____ involves the movement of particles down their concentration gradient through carrier proteins.

29. _____ ions are usually more concentrated inside a cell than outside the cell.

30. A(n) _____ _____ amplifies the communication from a signal molecule.

Read each question, and write your answer in the space provided.

31. Describe the electrical charge inside and outside a typical cell. Then explain how this affects an ion's ability to move into the cell.

_____

_____

_____

_____

_____

_____

32. Suppose you have to explain a concentration gradient to someone. Create a scenario that illustrates passive transport down the concentration gradient.

_____

_____

_____

_____

33. Name the three transport processes in cells that do not require energy, and briefly describe how each of them works.

_____

_____

_____

_____

_____

34. Using your understanding of osmosis, describe why putting salt on a pork chop before cooking it on a grill is likely to result in a dry, tough piece of meat.

_____

_____

_____

_____

35. How is facilitated diffusion different from the other passive transport processes?

_____

_____

_____

36. How does a cell consume a food particle that is too large to pass through a protein channel?

_____

_____

_____

Name _____ Date _____ Class _____

Cells and Their Environment

In the space provided, write the letter of the description that best matches the term or phrase.

_____ 1. passive transport

_____ 2. concentration gradient

_____ 3. equilibrium

_____ 4. diffusion

_____ 5. osmosis

_____ 6. hypertonic solution

_____ 7. hypotonic solution

_____ 8. isotonic solution

_____ 9. ion channel

_____ 10. carrier protein

_____ 11. facilitated diffusion

_____ 12. active transport

_____ 13. sodium-potassium pump

_____ 14. endocytosis

_____ 15. exocytosis

_____ 16. receptor protein

_____ 17. second messenger

a. movement of a substance down the substance's concentration gradient

b. causes a cell to shrink because of osmosis

c. movement of a substance by a vesicle to the outside of a cell

d. carrier protein used in active transport

e. protein used to transport specific substances

f. transport protein through which ions can pass

g. movement of a substance by a vesicle to the inside of a cell

h. does not require energy from the cell

i. concentration of a substance is equal throughout a space

j. difference in the concentration of a substance across a space

k. diffusion of water through a selectively permeable membrane

l. causes a cell to swell because of osmosis

m. passive transport using carrier proteins

n. produces no change in cell volume because of osmosis

o. movement of a substance against the substance's concentration gradient

p. acts as a signal molecule in the cytoplasm

q. binds to a signal molecule, enabling the cell to respond to the signal molecule

Name_____ Date _____ Class_____

Cells and Their Environment

Use the information below and the figure at right to answer questions 1–3.

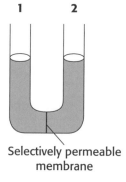

Selectively permeable membrane

Experiment A

A selectively permeable membrane separates the solutions in the arms of the U-tube shown at right. The membrane is permeable to water and to substance A but not to substance B. Forty grams of substance A and 20 g of substance B have been added to the water on side 1 of the U-tube. Twenty grams of substance A and 40 g of substance B have been added to the water on side 2 of the U-tube. Assume that after a period of time, the solutions on either side of the membrane have reached equilibrium.

Read each question, and write your answer in the space provided.

1. How many grams of substance A will be in solution on side 1 of the U-tube? How many grams of substance A will be in solution on side 2? Explain.

_____

_____

_____

_____

_____

2. How many grams of substance B will be in solution on side 1 of the U-tube? How many grams of substance B will be in solution on side 2? Explain.

_____

_____

_____

3. What has happened to the water level in the U-tube? Explain.

_____

_____

_____

_____

Use the information below to answer questions 4–6.

Experiment B

The cell membrane of red blood cells is permeable to water but not to sodium chloride, NaCl. Suppose that you have three flasks:

- Flask X contains a solution that is 0.5 percent NaCl.
- Flask Y contains a solution that is 0.9 percent NaCl.
- Flask Z contains a solution that is 1.5 percent NaCl.

To each flask, you add red blood cells, which contain a solution that is 0.9 percent NaCl.

Read each question, and write your answer in the space provided.

4. Predict what will happen to the red blood cells in flask X.

_____

_____

_____

_____

_____

5. Predict what will happen to the red blood cells in flask Y.

_____

_____

_____

_____

_____

6. Predict what will happen to the red blood cells in flask Z.

_____

_____

_____

_____

_____

Name_____ Date _____ Class_____

Photosynthesis and Cellular Respiration

In the space provided, write the letter of the term or phrase that best completes each statement or best answers each question.

_____ 1. Photosynthetic organisms get energy from
    a. inorganic substances.
    b. light.
    c. autotrophs.
    d. heterotrophs.

_____ 2. Which of the following correctly sequences the flow of energy?
    a. bacteria, fungus, rabbit
    b. bacteria, sun, flower, deer
    c. sun, grass, rabbit, fox
    d. sun, hawk, mouse

_____ 3. The production of ATP during photosynthesis requires
    a. energy released when hydrogen ions move down their concentration gradient.
    b. a carrier protein to catalyze the addition of a phosphate group to a molecule of ADP.
    c. energy from electrons passing through electron transport chains.
    d. All of the above

_____ 4. ATP molecules
    a. produce NADPH.
    b. contain five phosphate groups.
    c. can both store energy and provide it for metabolic reactions.
    d. help a plant produce carbon dioxide.

_____ 5. In glycolysis,
    a. aerobic processes occur.
    b. four ATP molecules are produced.
    c. four ADP molecules are produced.
    d. glucose is produced.

_____ 6. Carbon dioxide fixation in the Calvin cycle requires
    a. ATP and NADPH.
    b. ATP and $NADP^+$.
    c. ADP and NADPH.
    d. ATP and oxygen.

Question 7 refers to the chemical equation below.

$$3CO_2 + 3H_2O \xrightarrow{\text{light}} C_3H_6O_3 + 3O_2$$

_____ 7. This equation summarizes the overall process of
    a. cellular respiration.
    b. photosynthesis.
    c. the Calvin cycle.
    d. the Krebs cycle.

_____ 8. Which of the following environmental factors does NOT directly influence the rate of photosynthesis?
    a. light intensity
    b. oxygen concentration
    c. carbon dioxide concentration
    d. temperature

_____ 9. For each molecule of glucose entering glycolysis, there is a net gain of
   a. six ATP molecules.   c. three ATP molecules.
   b. four ATP molecules.   d. two ATP molecules.

_____ 10. Aerobic respiration follows glycolysis when _____ is available.
   a. carbon dioxide   c. hydrogen
   b. oxygen   d. water

_____ 11. If no oxygen is available to accept electrons during aerobic respiration,
   a. aerobic processes stop.
   b. fermentation proceeds.
   c. only small amounts of ATP can be produced.
   d. All of the above

_____ 12. Which of the following is NOT a fuel used for cellular respiration?
   a. carbohydrates   c. proteins
   b. fats   d. water

_____ 13. Electrons in pigment molecules become excited
   a. when light strikes a thylakoid.
   b. when water molecules are broken down.
   c. during light-independent reactions.
   d. during the Calvin cycle.

Question 14 refers to the chemical equation below.

$$C_6H_{12}O_6 + 6O_2 \xrightarrow{\text{enzymes}} 6CO_2 + 6H_2O + \text{energy}$$

_____ 14. The equation above summarizes the overall process of
   a. the Krebs cycle.   c. cellular respiration.
   b. photosynthesis.   d. the Calvin cycle.

_____ 15. During cellular respiration,
   a. the complete breakdown of glucose yields only carbon dioxide
      and water.
   b. the complete breakdown of glucose yields ATP molecules.
   c. NADPH is produced.
   d. carbon dioxide is required.

Question 16 refers to the figure below, which shows a chloroplast.

_____ 16. In the figure below, the reactions of the electron transport chains occur in
   the structure labeled
   a. *A.*   c. *C.*
   b. *B.*   d. *D.*

Circle **T** *if the statement is true or* **F** *if it is false.*

T F **17.** The reaction that removes a phosphate group from ATP and results in ADP provides energy for the cell.

T F **18.** Electrons and hydrogen ions combine with NADP$^+$ in an electron transport chain to produce NADPH.

T F **19.** The Calvin cycle produces ATP during the breakdown of a six-carbon molecule during cellular respiration.

T F **20.** In the third stage of photosynthesis, oxygen is used to make organic molecules.

T F **21.** Products of the Calvin cycle are three-carbon sugars that are used to produce organic compounds and regenerate the initial five-carbon compound.

T F **22.** Carotenoids are yellow and orange plant pigments.

T F **23.** The less carbon dioxide available to a plant, the faster photosynthesis proceeds.

T F **24.** The rate of photosynthesis increases as the concentration of hydrogen gas increases.

T F **25.** Metabolic processes that require oxygen are called anaerobic.

T F **26.** Fermentation allows the continued production of ATP even though oxygen is no longer present.

T F **27.** When glucose is broken down, two ATP molecules are used.

T F **28.** Some of the glucose required for cellular respiration in humans is obtained by eating cellulose.

Complete each statement by writing the correct term or phrase in the space provided.

29. Light-absorbing _____ are located in the membranes of

_____ .

30. The carrier protein that transports hydrogen ions across thylakoid membranes

and produce ATP acts as both a(n) _____ _____

and a(n) _____ .

31. The _____ _____ is the most common method of carbon dioxide fixation.

32. Aerobic respiration occurs in the _____ of eukaryotic cells.

33. Plants use sugars produced during _____ to make organic compounds.

34. During photosynthesis, light energy is converted to _____ energy.

35. During anaerobic processes, NADH transfers electrons to the pyruvate

produced during _____ .

36. Glycolysis is a biochemical pathway that breaks down a six-carbon glucose

molecule to two three-carbon _____ .

37. During aerobic respiration, pyruvate is first converted to acetyl-CoA, which

enters the _____ _____ .

38. During cellular respiration, a cell produces most of its energy through

_____ respiration.

• • • • • • • • • • • • • • •
Read each question, and write your answer in the space provided.

39. Explain how ATP provides energy for cells.

_____

_____

_____

40. Briefly explain how ATP is produced by electron transport chains during
photosynthesis.

_____

_____

_____

_____

_____

_____

41. Briefly describe how environmental factors affect the rate of photosynthesis.

_____

_____

_____

_____

42. Explain the benefits and uses of lactic acid fermentation and alcoholic
fermentation.

_____

_____

_____

_____

Name _____ Date _____ Class _____

Photosynthesis and Cellular Respiration

Write the correct term from the list below in the space next to its definition.

aerobic	cellular respiration	heterotrophs
anaerobic	chlorophyll	Krebs cycle
autotrophs	electron transport chain	photosynthesis
Calvin cycle	fermentation	pigment
carbon dioxide fixation	glycolysis	thylakoids
carotenoids		

_____ **1.** the process by which light energy is converted to chemical energy

_____ **2.** organisms that use energy from sunlight or inorganic substances to make organic compounds

_____ **3.** organisms that get energy by consuming food

_____ **4.** the process by which cells harvest energy from food

_____ **5.** a substance that absorbs light

_____ **6.** the primary pigment involved in photosynthesis

_____ **7.** absorb wavelengths of light different from those absorbed by chlorophyll

_____ **8.** the series of molecules down which excited electrons are passed in a thylakoid membrane

_____ **9.** the transfer of carbon dioxide to organic compounds

_____ **10.** a series of enzyme-assisted chemical reactions that produces a three-carbon sugar molecule

_____ **11.** a process that requires oxygen

_____ **12.** a process that does not require oxygen

_____ **13.** the process by which glucose is broken down to pyruvate

_____ **14.** a series of enzyme-assisted chemical reactions following glycolysis that produces carbon dioxide

_____ **15.** the recycling of NAD^+ under anaerobic conditions

_____ **16.** disk-shaped structures inside chloroplasts

CHAPTER

5   **SCIENCE SKILLS: INTERPRETING DATA**

Photosynthesis and Cellular Respiration

Scientists estimate that only 10 percent of the energy present at each level of the food chain is available to the next level. Scientists also estimate that only 1 percent of the light energy from the sun that reaches photosynthetic organisms is converted to chemical energy during photosynthesis. Use this information and the two food chains below to answer questions 1–5.

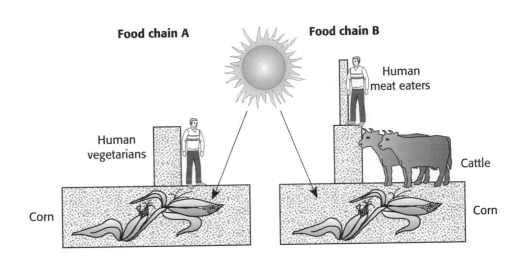

Food chain A   **Food chain B**

Human meat eaters

Human vegetarians

Cattle

Corn   Corn

Read each question, and write your answer in the space provided

1. Assume that 1 million kilocalories (kcal) of energy from the sun is available to the autotrophs in the food chains above. Determine the amount of energy that will be available at every level of each food chain.

_____

_____

_____

_____

2. Explain why most food chains consist of no more than three or four levels.

_____

_____

_____

_____

3. What happens to the stored energy that does not advance from one level of the food chain to the next?

_____

_____

_____

_____

_____

_____

4. A 55 kg person at rest requires an average of 54 kcal/hour. The same person engaged in activity requires an average of 87.5 kcal/hour. What is the percent increase in required kcal/hour between rest and activity? Which would be a more efficient diet for an active person, plant foods or meat? Explain.

_____

_____

_____

_____

_____

5. Earth's population is increasing at a rate that may outpace our ability to produce enough food. Why are some people promoting vegetarianism as an answer to this dilemma?

_____

_____

_____

_____

_____

_____

CHAPTER

6 **TEST PREP PRETEST**

Chromosomes and Cell Reproduction

In the space provided, write the letter of the term or phrase that best completes each statement or best answers each question.

_____ 1. As a cell prepares to divide, a DNA molecule and its associated proteins coil to form a

    a. chromatid.   **c.** chromosome.
    b. gene.   **d.** centromere.

_____ 2. The number of chromosomes found in a human body cell is

    a. 23.   **c.** 48.
    b. 46.   **d.** 64.

_____ 3. The condition in which a diploid cell has an extra chromosome is called

    a. monosomy.   **c.** trisomy.
    b. disjunction.   **d.** karyotype.

_____ 4. The sex of a human offspring is determined by

    a. the female.
    b. the male.
    c. both the female and the male.
    d. neither the female nor the male.

_____ 5. Bacteria reproduce through an asexual process called

    a. meiosis.   **c.** mitosis.
    b. cytokinesis.   **d.** binary fission.

_____ 6. The repeating sequence of growth and division through which many eukaryotic cells pass is called

    a. the cell cycle.   **c.** cytokinesis.
    b. binary fission.   **d.** meiosis.

_____ 7. Which of the following is NOT a stage of mitosis?

    a. prophase   **c.** metaphase
    b. synthesis   **d.** telophase

_____ 8. In human sexual reproduction, a male haploid gamete and a female haploid gamete unite to form a(n)

    a. egg cell with 46 chromosomes.
    b. zygote with 23 chromosomes.
    c. zygote with 46 chromosomes.
    d. sperm cell with 23 chromosomes.

_____ 9. In plant cells, cytokinesis requires the formation of a new

    a. cell membrane.   **c.** centromere.
    b. cell wall.   **d.** series of protein threads.

_____ 10. Normal cells can become cancer cells through

    a. gene mutations.   **c.** viruses.
    b. ultraviolet radiation.   **d.** All of the above

In the space provided, write the letter of the description that
best matches the term or phrase.

_____ 11. chromatids

_____ 12. centromere

_____ 13. homologous chromosomes

_____ 14. diploid

_____ 15. gametes

_____ 16. haploid

_____ 17. zygote

_____ 18. DNA

_____ 19. karyotype

_____ 20. cytokinesis

_____ 21. mutations

_____ 22. deletion

_____ 23. cell growth (G_1) checkpoint

_____ 24. anaphase

_____ 25. prophase

a. egg cells and sperm cells

b. a mutation that occurs when a
chromosome fragment breaks off
and is lost

c. cytoplasm divides

d. two chromatids move to opposite poles
of the cell

e. supplies information that directs a cell's
activities and determines its
characteristics

f. the point at which two chromatids are
attached

g. determines whether a cell will divide

h. the two copies of DNA on each
chromosome that form just before cell
division

i. a cell that contains one set of
chromosomes

j. chromosomes that are similar in shape
and size and have similar genetic
information

k. the nuclear membrane dissolves and
chromosomes become visible

l. changes in an organism's genetic
material

m. a diploid cell that results from the
fusion of two haploid gametes

n. a cell that contains two sets of
chromosomes

o. a picture of the chromosomes found in
an individual's cells

Complete each statement by writing the correct term or phrase
in the space provided.

26. A(n) _____ is a segment of DNA that transmits information
from parent to offspring.

27. An individual with an extra copy of chromosome 21 demonstrates traits

collectively known as _____ _____ .

28. The 22 pairs of chromosomes in human somatic cells that are the same in males

and females are called _____ .

29. The human chromosomes that determine an individual's sex are called the

_____ _____ .

Questions 30–32 refer to the sequence below.

$$G_1 \longrightarrow S \longrightarrow G_2 \longrightarrow M \longrightarrow C$$

30. The sequence above represents the _____

_____ .

31. The C in the sequence represents the phase in which _____
occurs.

32. Phases G_1, S, and G_2 in the sequence above are collectively called

_____ .

33. Each individual protein structure that helps to move the chromosomes apart

during mitosis is called a(n) _____ .

34. _____ is a disease caused by uncontrolled cell division.

35. In the first stage of binary fission, the DNA is _____ .

36. The microtubules that form centrioles and spindle fibers are made of hollow

tubes of _____ .

Questions 37–40 refer to the figures below.

Animal cell

Plant cell

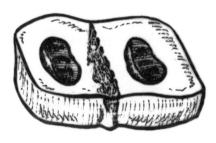

37. The process shown in the figures above is _____ .

38. The structure in the center of the animal cell that pinches the cell in half is

composed of _____ _____ .

39. The process shown above occurs at the end of _____ .

40. The end result of the process shown above is the formation of two new

_____ .

Read each question, and write your answer in the space provided.

41. What happens to the structure of DNA in your cells prior to cell division?

_____

_____

_____

42. Explain the difference in the number of chromosomes between a frog somatic cell and a frog egg cell.

_____

_____

43. What happens when nondisjunction takes place during cell division?

_____

_____

_____

_____

44. Briefly describe what happens at each checkpoint during the cell cycle.

_____

_____

_____

_____

_____

45. Briefly explain the five phases of the cell cycle.

_____

_____

_____

46. What are the four stages of mitosis, in the correct order?

a. _____ c. _____

b. _____ d. _____

47. Explain the importance of mitosis and cytokinesis to multicellular organisms.

_____

_____

Name_____ Date _____ Class _____

Chromosomes and Cell Reproduction

In the space provided, write the letter of the term or phrase that best completes each statement or best answers each question.

_____ 1. An organism's reproductive cells, such as sperm or egg cells, are called
   a. genes.
   b. chromosomes.
   c. gametes.
   d. zygotes.

_____ 2. A form of asexual reproduction in bacteria is
   a. binary fission.
   b. trisomy.
   c. mitosis.
   d. development.

_____ 3. A segment of DNA that codes for a protein or RNA molecule is a
   a. chromosome.
   b. gene.
   c. chromatid.
   d. centromere.

_____ 4. At the beginning of cell division, DNA and the proteins associated with the DNA coil into a structure called a(n)
   a. chromatid.
   b. autosome.
   c. centromere.
   d. chromosome.

_____ 5. The two exact copies of DNA that make up each chromosome are called
   a. homologous chromosomes.
   b. centromeres.
   c. chromatids.
   d. autosomes.

_____ 6. The two chromatids of a chromosome are attached at a point called the
   a. diploid.
   b. centriole.
   c. spindle.
   d. centromere.

_____ 7. Chromosomes that are similar in size, shape, and genetic content are called
   a. homologous chromosomes.
   b. haploid.
   c. diploid.
   d. karyotypes.

_____ 8. When a cell contains two sets of chromosomes, it is said to be
   a. haploid.
   b. binary.
   c. diploid.
   d. saturated.

_____ 9. When a cell contains one set of chromosomes, it is said to be
   a. haploid.
   b. separated.
   c. diploid.
   d. homologous.

_____ 10. The fertilized egg, the first cell of a new individual, is called a(n)
   a. autosome.
   b. zygote.
   c. organism.
   d. chromosome.

_____ 11. A photo of the chromosomes in a dividing cell, arranged by size, is a(n)
   a. electronic scan.
   b. karyotype.
   c. X ray.
   d. anaphase.

(continued on next page)

_____ **12.** Chromosomes not directly involved in determining the sex of an individual are called

    a. asexual chromosomes.   **c.** autosomes.
    b. chromatids.   **d.** haploid.

_____ **13.** Chromosomes that contain genes that will determine the sex of the individual are called

    a. X chromosomes.   **c.** Y chromosomes.
    b. sex chromosomes.   **d.** autosomes.

_____ **14.** The repeated sequence of growth and division during the life of a cell is called the

    a. cell cycle.   **c.** binary fission.
    b. cytokinesis.   **d.** amniocentesis.

_____ **15.** The first three phases of the cell cycle are called

    a. anaphase.   **c.** mitosis.
    b. interphase.   **d.** synthesis phase.

_____ **16.** The process during which the nucleus of a cell is divided into two nuclei is called

    a. fertilization.   **c.** mitosis.
    b. disjunction.   **d.** cytokinesis.

_____ **17.** The process during cell division in which the cytoplasm divides is called

    a. cytokinesis.   **c.** interphase.
    b. trisomy.   **d.** mitosis.

_____ **18.** The uncontrolled division of cells is called

    a. Down syndrome.   **c.** cancer.
    b. mutation.   **d.** trisomy.

_____ **19.** Cell structures made of individual microtubule fibers that are involved in moving chromosomes during cell division are called

    a. chromatids.   **c.** centrioles.
    b. fertilizers.   **d.** spindles.

Chromosomes and Cell Reproduction

The figure below illustrates the life cycle of a eukaryotic cell, which is known as the cell cycle. The names of the phases have been omitted from the figure; however, they are listed in random order below (a–h). Use the figure below to complete items 1–8.

In the space provided in the figure below, write the letter of the phase of the cell cycle that matches each phase in the figure.

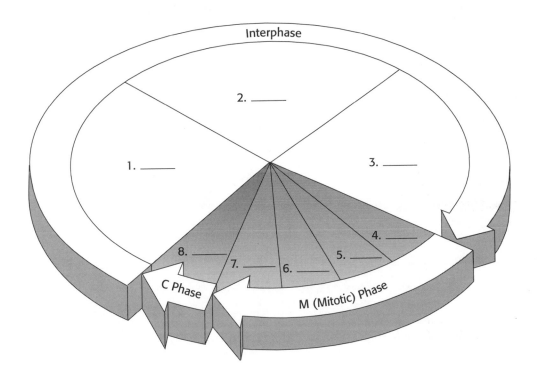

Phases of the Cell Cycle

a. prophase

b. G_1

c. telophase

d. metaphase

e. S

f. cytokinesis

g. G_2

h. anaphase

(continued on next page)

Each animal cell in the figure below shows a different stage of mitosis. Use the figure below to complete items 9–16.

In the space provided below each cell in the figure, write the name of the stage of mitosis that is represented.

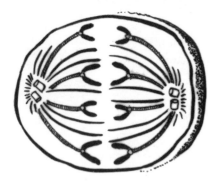

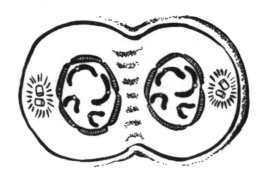

9. _____

10. _____

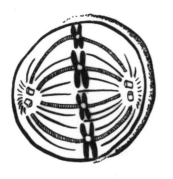

11. _____

12. _____

Determine the order in which the following four stages of mitosis take place. Write the number of each step (1–4) in the space provided.

_____ 13. anaphase

_____ 14. metaphase

_____ 15. telophase

_____ 16. prophase

Name _____ Date _____ Class _____

Meiosis and Sexual Reproduction

In the space provided, write the letter of the term or phrase that best completes each statement or best answers each question.

_____ 1. An advantage of sexual reproduction is that
  a. many offspring are produced in a short time.
  b. it increases genetic diversity.
  c. production of gametes requires energy.
  d. organisms remain stable in a changing environment.

_____ 2. Crossing-over occurs
  a. during prophase II.   **c.** during prophase I.
  b. during fertilization.   **d.** at the centromere.

_____ 3. Cytoplasm divides unequally in meiosis during production of
  a. gametes.   **c.** cytokinesis.
  b. sperm cells.   **d.** egg cells.

_____ 4. The zygote is the only diploid cell in
  a. the haploid life cycle.   **c.** the diploid life cycle.
  b. asexual reproduction.   **d.** animals.

_____ 5. Which of the following does NOT provide new genetic combinations?
  a. random fertilization   **c.** independent assortment
  b. cytokinesis   **d.** crossing-over

_____ 6. Which of the following is NOT a type of asexual reproduction?
  a. budding   **c.** fission
  b. fragmentation   **d.** alternation of generations

_____ 7. If, during an animal's life cycle, the gametes are the only haploid cells, the life cycle is
  a. alternation of generations. **c.** a diploid life cycle.
  b. a haploid life cycle.   **d.** mutated.

_____ 8. DNA replication occurs
  a. after telophase I.
  b. prior to prophase I.
  c. in both meiosis I and meiosis II.
  d. when the chromosomes align at the cell's equator.

_____ 9. In telophase II, cytokinesis results in
  a. two haploid cells.   **c.** four haploid cells.
  b. two diploid cells.   **d.** four diploid cells.

_____ 10. Asexual reproduction produces
  a. clones.   **c.** gametophytes.
  b. spores.   **d.** polar bodies.

In the space provided, write the letter of the description that best matches the term or phrase.

_____ 11. gametophyte

_____ 12. crossing-over

_____ 13. life cycle

_____ 14. clone

_____ 15. independent assortment

_____ 16. spore

_____ 17. spermatogenesis

_____ 18. sporophyte

_____ 19. polar body

_____ 20. fragmentation

_____ 21. oogenesis

_____ 22. budding

_____ 23. anaphase I

_____ 24. fertilization

_____ 25. asexual reproduction

_____ 26. ovum

a. random distribution of homologous chromosomes during meiosis

b. multicellular, haploid phase in alternation of generations

c. a method of asexual reproduction in which the body breaks in several pieces

d. produces spores in the diploid phase of a plant's life cycle

e. small cell with very little cytoplasm that is formed during oogenesis and eventually dies

f. all copies of the single parent's genes are passed to the offspring

g. portions of a chromatid on one homologous chromosome break off and trade places with the corresponding portion on one of the chromatids of the other homologous chromosome

h. the process by which gametes are produced in male animals

i. new individuals split off from existing ones

j. the union of sperm and egg cells to produce a diploid zygote

k. the activities in the life of an organism from one generation to the next

l. haploid reproductive cell of plants

m. offspring that is genetically identical to its parent

n. female gamete, also called an egg

o. the process by which gametes are produced in female animals

p. homologous chromosomes move to opposite poles of the cell

Complete each statement by writing the correct term or phrase in the space provided.

27. Asexual reproduction limits _____ diversity.

28. Spermatogenesis produces _____ sperm cells.

29. _____ _____ , although not part of meiosis, increases the number of possible genetic combinations.

30. Asexual reproduction methods include _____ , fragmentation, and _____ .

31. In the haploid life cycle, gametes are produced by _____ , and the zygote is produced by _____ .

32. When corresponding portions of chromatids on two homologous chromosomes change places, _____ _____ has occurred.

33. Only one ovum is produced by _____ .

34. In plants that have alternation of generations, the haploid _____ produces the gametes.

35. Increased genetic variation often increases the rate of _____ .

36. Meiosis in plants often produces _____ , haploid cells that later lead to the production of gametes.

37. Crossing-over is an efficient way to produce _____ _____ , which increases genetic diversity.

Questions 38–41 refer to the figure below.

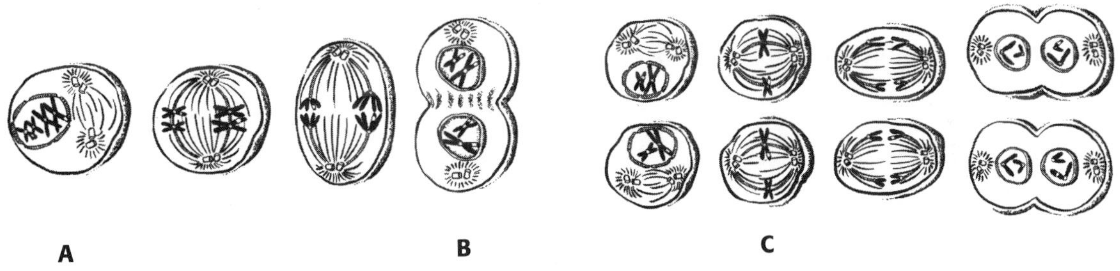

A   **B**   **C**

38. The process shown above is called _____ .

39. In the process shown above, label *A* refers to _____ .

40. In the process shown above, label *B* refers to _____ and _____ .

41. In the process shown above, label *C* refers to _____ .

• • • • • • • • • • • • •
Read each question, and write your answer in the space provided.

42. Describe the similarities and differences between the formation of male and female gametes in animals.

_____

_____

_____

_____

43. Identify and describe the three types of asexual reproduction.

_____

_____

_____

_____

44. What is the difference between anaphase I and anaphase II? Why is the difference significant?

_____

_____

_____

_____

_____

45. Describe the haploid and diploid life cycles.

_____

_____

_____

_____

_____

46. Explain what happens during alternation of generations in plants.

_____

_____

_____

_____

_____

47. How does crossing-over affect evolution?

_____

_____

_____

_____

Meiosis and Sexual Reproduction

Complete the crossword puzzle using the clues provided.

ACROSS

6. gamete-producing process that occurs in male reproductive organs

7. the kind of reproduction in which two parents form haploid cells that join to produce offspring

9. female gamete

11. occurs during prophase I of meiosis

12. a haploid plant reproductive cell produced by meiosis

13. form of cell division that halves the number of chromosomes

14. the type of assortment that involves the random distribution of homologous chromosomes during meiosis

DOWN

1. the haploid phase of a plant that produces gametes by mitosis

2. the process in most animals that produces diploid zygotes

3. an individual produced by asexual reproduction

4. the kind of reproduction in which a single parent passes copies of all its genes to its offspring

5. the name for the cycle that spans from one generation to the next

7. male gametes

8. diploid phase of a plant that produces spores

10. occurs in the ovaries

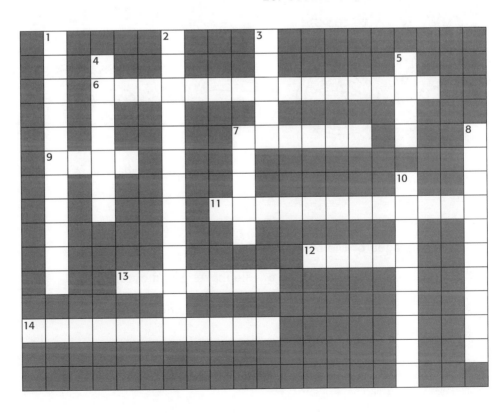

CHAPTER
7

SCIENCE SKILLS: SEQUENCING/ORGANIZING INFORMATION

Meiosis and Sexual Reproduction

Examine the figure below, which shows the stages of meiosis. Use the figure below to complete items 1–8.

In the space provided in the figure below, write the letter of the stage of meiosis from the list below (a–h) that matches each stage in the figure.

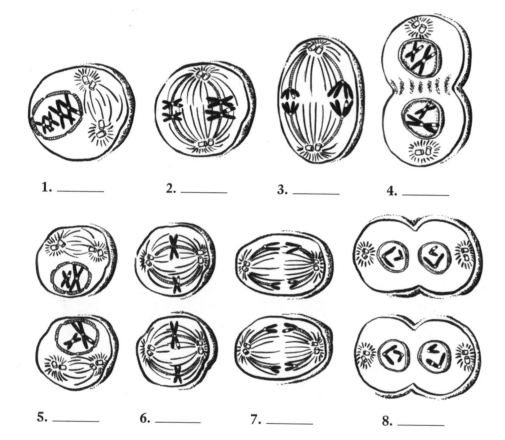

1. _____   2. _____   3. _____   4. _____

5. _____   6. _____   7. _____   8. _____

Stages of Meiosis

a. anaphase II

b. metaphase I

c. anaphase I

d. metaphase II

e. telophase II and cytokinesis

f. telophase I and cytokinesis

g. prophase I

h. prophase II

In the space provided, write the letter of the description that best matches the stage of meiosis.

_____ 9. metaphase I

_____ 10. prophase II

_____ 11. telophase I

_____ 12. metaphase II

_____ 13. telophase II

_____ 14. anaphase II

_____ 15. prophase I

_____ 16. anaphase I

a. A new spindle forms around the chromosomes.

b. Chromatids remain attached at their centromeres as the spindle fibers move the homologous chromosomes to opposite poles of the cell.

c. A nuclear envelope forms around each set of chromosomes, the spindle breaks down, and the cytoplasm divides, resulting in four haploid cells.

d. Chromosomes gather at the poles; the cytoplasm divides.

e. The nuclear envelope breaks down; genetic material is exchanged through crossing-over.

f. Chromosomes line up at the equator.

g. Pairs of homologous chromosomes line up at the equator.

h. Centromeres divide, enabling the chromatids, now called chromosomes, to move to opposite poles of the cell.

Name_____ Date _____ Class _____

CHAPTER
8 **TEST PREP PRETEST**

Mendel and Heredity

In the space provided, write the letter of the term or phrase that best completes each statement or best answers each question.

_____ 1. *Pisum sativum*, the garden pea, is a good subject to use in studying heredity for all of the following reasons EXCEPT
 a. Several varieties of *Pisum sativum* are available that differ in easily distinguishable traits.
 b. *Pisum sativum* is a small, easy-to-grow plant.
 c. *Pisum sativum* matures quickly and produces a large number of offspring.
 d. A *Pisum sativum* plant with male reproductive parts must cross-pollinate with a plant having female reproductive parts for reproduction to take place.

_____ 2. Step 1 of Mendel's garden pea experiment, allowing each variety of garden pea to self-pollinate for several generations, produced the
 a. F_1 generation.   c. P generation.
 b. F_2 generation.   d. P_2 generation.

_____ 3. In the F_2 generation in Mendel's experiments, the ratio of dominant to recessive phenotypes was
 a. 1:3.   c. 2:1.
 b. 1:2.   d. 3:1.

_____ 4. The trait that was expressed in the F_1 generation in Mendel's experiment is considered
 a. recessive.   c. second filial.
 b. dominant.   d. parental.

_____ 5. Mendel's law of segregation states that
 a. pairs of alleles are dependent on one another when separation occurs during gamete formation.
 b. pairs of alleles separate independently of one another after gamete formation.
 c. each pair of alleles remains together when gametes are formed.
 d. the two alleles for a trait separate when gametes are formed.

_____ 6. A series of genetic crosses results in 787 long-stemmed plants and 277 short-stemmed plants. The probability that you will obtain short-stemmed plants if you repeat this experiment is
 a. $\frac{277}{1,064}$.   c. $\frac{787}{277}$.
 b. $\frac{277}{787}$.   d. $\frac{787}{1,064}$.

_____ 7. Crossing a snapdragon that has red flowers with one that has white flowers produces a snapdragon that has pink flowers. The trait for flower color exhibits
 a. multiple alleles.   c. incomplete dominance.
 b. complete dominance.   d. codominance.

_____ 8. Which of the following is NOT considered a genetic disorder?
   a. sickle cell anemia
   b. hemophilia
   c. AIDS
   d. cystic fibrosis

_____ 9. On which of the following chromosomes would a sex-linked trait most likely be found in humans?
   a. X
   b. Y
   c. O
   d. YO

_____ 10. The roan color of a horse is an example of
   a. homozygous alleles.
   b. codominance.
   c. incomplete dominance.
   d. Both (a) and (c)

Questions 11 and 12 refer to the figure at right, which represents a monohybrid cross between two individuals that are heterozygous for a trait.

	D	d
D	DD	Dd
__	D__	d__

_____ 11. If the resulting phenotypic ratio is 3:1, the missing parental allele is
   a. *d.*
   b. *D.*
   c. *Dd.*
   d. *DD.*

_____ 12. The two unknown genotypes in the offspring are
   a. *DD* and *dd.*
   b. *Dd* and *Dd.*
   c. *dd* and *DD.*
   d. *Dd* and *dd.*

• • • • • • • • • • • • • •

Circle T *if the statement is true or* F *if it is false.*

T F 13. The transmission of traits from parents to offspring is called heredity.

T F 14. When Mendel cross-pollinated two varieties from the P generation that exhibited contrasting traits, he called the offspring the second filial, or F_2, generation.

T F 15. In Mendel's experiments, the recessive traits reappeared in the F_2 generation in approximately 25 percent of the plants.

T F 16. What Mendel called factors are today called genes.

T F 17. Each version of a gene is called an allele.

T F 18. Mendel's law of independent assortment holds true for all genes, regardless of location.

T F 19. When you toss a coin, the probability that it will land heads up is $\frac{1}{4}$, or 25 percent.

T F 20. In a test cross, one individual has a dominant phenotype and unknown genotype and the other is homozygous recessive.

T F 21. Genes with more than two alleles are referred to as having multiple alleles.

T F 22. Hemophilia is caused by a mutated allele that produces a defective form of the protein hemoglobin.

Questions 23 and 24 refer to the figure below, which shows the inheritance of sickle cell anemia in a family.

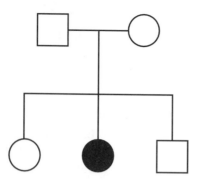

T   F   **23.** The figure above is referred to as a pedigree.

T   F   **24.** Both parents in this diagram carry a recessive allele for sickle cell anemia but do not exhibit the symptoms of this disorder.

• • • • • • • • • • • • • •

Complete each statement by writing the correct term or phrase in the space provided.

25. The study of heredity is called _____ .

26. A flower fertilizes itself through a process called _____

_____ .

27. The _____ , or physical appearance, of an individual is determined by the alleles that code for traits. The set of alleles that an

individual has is called its _____ .

28. The likelihood that a specific event will occur is called _____ .

29. A genetic cross can be represented graphically in the form of a(n)

_____ _____ .

30. A cross between a pea plant that is true breeding for green pod color and one that is true breeding for yellow pod color is an example of a(n)

_____ cross.

31. Characteristics such as eye color, height, weight, hair and skin color are

examples of _____ _____ because several genes act together to influence a trait.

32. Mutations in genetic material may cause _____

_____ , such as cystic fibrosis and muscular dystrophy.

33. _____ is a condition that impairs the blood's ability to clot.

34. _____ , a genetic disorder that can cause severe mental retardation results because an enzyme that converts one amino acid into another is missing.

Read each question, and write your answer in the space provided.

35. In step 1 of his experiments, how did Mendel ensure that each variety of garden pea was true breeding for a particular trait?

_____

_____

36. What approximate ratio of plants expressing contrasting traits did Mendel calculate in his F_2 generation of garden peas? What steps did he take to calculate this ratio?

_____

_____

_____

37. Name Mendel's two laws of heredity.

_____

_____

38. Give an example of how the environment might influence gene expression.

_____

_____

_____

39. Describe how persons who are heterozygous for sickle cell anemia are protected against malaria.

_____

_____

_____

40. Describe one sex-linked genetic disorder.

_____

_____

Name_____ Date _____ Class _____

Mendel and Heredity

In the space provided, write the letter of the description that best matches the term or phrase.

_____ 1. heredity

_____ 2. genetics

_____ 3. monohybrid cross

_____ 4. true breeding

_____ 5. P generation

_____ 6. F_1 generation

_____ 7. F_2 generation

_____ 8. alleles

_____ 9. dominant

_____ 10. recessive

_____ 11. homozygous

_____ 12. heterozygous

_____ 13. genotype

_____ 14. phenotype

_____ 15. law of segregation

_____ 16. law of independent assortment

a. the alleles of a particular gene are different

b. the two alleles for a trait separate when gametes are formed

c. the alleles of different genes separate independently of one another during gamete formation

d. not expressed when the dominant form of the trait is present

e. passing of traits from parents to offspring

f. all the offspring display only one form of a particular trait

g. the expressed form of a trait

h. first two individuals crossed in a breeding experiment

i. physical appearance of a trait

j. a cross that considers one pair of contrasting traits

k. offspring of the F_1 generation

l. when the two alleles of a particular gene are the same

m. branch of biology that studies heredity

n. different versions of a gene

o. offspring of the P generation

p. set of alleles that an individual has

Write the correct term from the list below in the space next to its definition.

codominance   pedigree   Punnett square

incomplete dominance   polygenic trait   sex-linked trait

multiple alleles   probability   test cross

_____ 17. diagram that predicts the outcomes of a genetic cross

_____ 18. cross of a homozygous recessive individual with an individual with a dominant phenotype of unknown genotype

_____ 19. the likelihood that a specific event will occur

_____ 20. a family history that shows how a trait is inherited

_____ 21. trait whose allele is located on the X chromosome

_____ 22. when several genes influence a trait

_____ 23. when an individual displays a trait that is intermediate between the two parents

_____ 24. two dominant alleles are expressed at the same time

_____ 25. genes with three or more alleles

Name_____ Date_____ Class_____

Mendel and Heredity

Part of Gregor Mendel's hypotheses of inheritance proposed that when two contrasting traits occur together, one of them may be completely expressed, while the other may have no observable effect on the organism's appearance. He called the expressed trait dominant and the unexpressed trait recessive. In his genetic experiments, Mendel studied seven contrasting traits of peas. One of the traits he studied was the trait for seed shape. Mendel found that there are two forms of the trait for seed shape—wrinkled and round. In an experiment to determine which form of this trait was dominant, Mendel performed the following experiment:

A. Mendel allowed the plants to self-fertilize to produce the homozygous P generations.

B. Mendel then performed a cross between the homozygous plants with round seeds (*RR*) and homozygous plants with wrinkled seeds (*rr*) to produce the F_1 generation.

C. Mendel then allowed the F_1 plants to self-fertilize to produce the F_2 generation.

D. Mendel analyzed the results of the crosses to determine the genotypes and phenotypes present.

Use the information above to answer questions 1–5.

1. To determine the results of this experiment, complete the Punnett squares below by writing the correct allele(s) in the space provided.

P RR ___ × ___ rr

F_1 a. _____ ▼ b. _____

    c. _____ | d. _____ | e. _____
              | |
    f. _____ | g. _____ | h. _____

F_2 i. _____ ▼ j. _____

    k. _____ | l. _____ | m. _____
              | |
    n. _____ | o. _____ | p. _____

Read each question, and write your answer in the space provided.

2. What phenotypes are present in the F_1 generation? What genotypes are present?

_____

_____

_____

_____

3. What phenotypes are present in the F_2 generation? In what ratio are they present?

_____

_____

_____

_____

4. Does your analysis support or refute Mendel's hypothesis of dominant and recessive inheritance? Explain.

_____

_____

_____

_____

_____

5. Do you think the ratios of F_2 phenotypes that Mendel observed in his experiment were exactly the same as the F_2 phenotype ratio that you calculated? Explain.

_____

_____

_____

_____

_____

Name_____ Date _____ Class_____

DNA: The Genetic Material

In the space provided, write the letter of the term or phrase that best completes each statement or best answers each question.

_____ 1. In 1928, the experiments of Frederick Griffith demonstrated transformation of
   a. *R* bacteria into *S* bacteria.
   b. *S* bacteria into *R* bacteria.
   c. heat-killed *S* bacteria into *R* bacteria.
   d. *S* bacteria into heat-killed *R* bacteria.

_____ 2. In 1952, Alfred Hershey and Martha Chase used the bacteriophage T2 and radioactive labels to determine that genetic material is made of
   a. protein.
   b. RNA.
   c. DNA.
   d. ^{35}S.

_____ 3. Each nucleotide in a DNA molecule consists of a
   a. sulfur group, a five-carbon sugar molecule, and a nitrogen base.
   b. phosphate group, a six-carbon sugar molecule, and a nitrogen base.
   c. phosphate group, a five-carbon sugar molecule, and an oxygen base.
   d. phosphate group, a five-carbon sugar molecule, and a nitrogen base.

_____ 4. In 1953, James Watson and Francis Crick built a model of DNA with the configuration of a
   a. single helix.
   b. double helix.
   c. triple helix.
   d. circle.

_____ 5. DNA is replicated before
   a. crossing-over.
   b. cell division.
   c. cell death.
   d. the G_1 phase.

_____ 6. Until 1952, many scientists thought genetic material was composed of protein because
   a. protein had been identified in bacteria.
   b. proteins were involved in so many other aspects of cell function.
   c. it was known that proteins could replicate.
   d. Mendel had isolated proteins in his heredity experiments.

7. Purines and pyrimidines differ in that
   a. purines have a single ring of carbon and nitrogen, whereas pyrimidines have a double ring.
   b. purines have no nitrogen in their molecular structure.
   c. pyrimidines have no nitrogen in their molecular structure.
   d. purines have a double ring of carbon and nitrogen, whereas pyrimidines have a single ring.

8. The work of Chargaff, Wilkins, and Franklin formed the basis for
   a. Watson and Crick's DNA model.
   b. Hershey and Chase's work on bacteriophages.
   c. Avery's work on transformation.
   d. Griffith's discovery of transformation.

9. At the end of the replication process, each of the two new DNA molecules is composed of
   a. two new DNA strands.
   b. one new and one original DNA strand.
   c. one new and one mutated DNA strand.
   d. two original DNA strands.

• • • • • • • • • • • • • • •
In the space provided, write the letter of the description that best matches the term or phrase.

_____ 10. vaccine

_____ 11. bacteriophage

_____ 12. nucleotide

_____ 13. deoxyribose

_____ 14. adenine

_____ 15. guanine

_____ 16. cytosine

_____ 17. thymine

_____ 18. purines

_____ 19. pyrimidines

_____ 20. helicases

_____ 21. DNA polymerases

_____ 22. replication forks

_____ 23. eukaryotic DNA

_____ 24. bacterial DNA

a. a nitrogen base that forms hydrogen bonds with cytosine

b. a virus that infects bacteria

c. a long DNA molecule

d. a substance that is introduced into the body to produce immunity

e. enzymes that open up the double helix by breaking the hydrogen bonds that link complementary bases

f. a class of organic molecules, each having a single ring of carbon and nitrogen atoms

g. a circular DNA molecule

h. a nitrogen base that forms hydrogen bonds with guanine

i. a nitrogen base that forms hydrogen bonds with thymine

j. enzymes that move along each of the DNA strands during replication, adding nucleotides to the exposed bases

k. a class of organic molecules, each having a double ring of carbon and nitrogen atoms

l. portions of DNA where the double helix separates during DNA replication

m. a five-carbon sugar

n. consists of a phosphate group, a sugar molecule, and a nitrogen base

o. a nitrogen base that forms hydrogen bonds with adenine

Complete each statement by writing the correct term or phrase in the space provided.

25. In 1928, Frederick Griffith found that the capsule that enclosed one strain of

    Streptococcus pneumoniae caused the microorganism's _____ .

26. Avery's experiments demonstrated that DNA is the _____
    material.

27. After infecting *Escherichia coli* bacteria with ^{32}P-labeled phages, Hershey and
    Chase traced the ^{32}P. The scientists found most of the radioactive substance in

    the _____ .

28. The circular DNA molecules in prokaryotes usually contain

    _____ replication forks during replication, while linear
    eukaryotic DNA contains many more.

29. Chargaff's observations established the _____

    _____ rules, which describe the specific pairing between bases
    on DNA strands.

30. Watson and Crick used the X-ray _____ photographs of
    Wilkins and Franklin to build their model of DNA.

31. The strict arrangement of base-pairings in the double helix results in two

    strands of nucleotides that are _____ to each other.

32. The process of making new DNA is called _____ .

33. The point at which the double helix separates during replication is called

    the _____ _____ .

34. DNA replication occurs during the _____ phase of the
    cell cycle.

35. Errors made during the replication process are corrected by DNA polymerase's

    ability to _____ the new DNA strand.

Read each question, and write your answer in the space provided.

36. What happened when Griffith mixed harmless living *R* bacteria with harmless
    heat-killed *S* bacteria and then injected mice with this mixture?

    _____

    _____

    _____

    _____

37. How did Avery's experiment identify the material responsible for transformation?

_____

_____

_____

38. Why did Hershey and Chase use radioactive elements in their experiments?

_____

_____

_____

39. Explain how DNA polymerase "proofreads" a new DNA strand.

_____

_____

_____

_____

40. Describe the role of DNA helicases during replication.

_____

_____

_____

Questions 41 and 42 refer to the figure below.

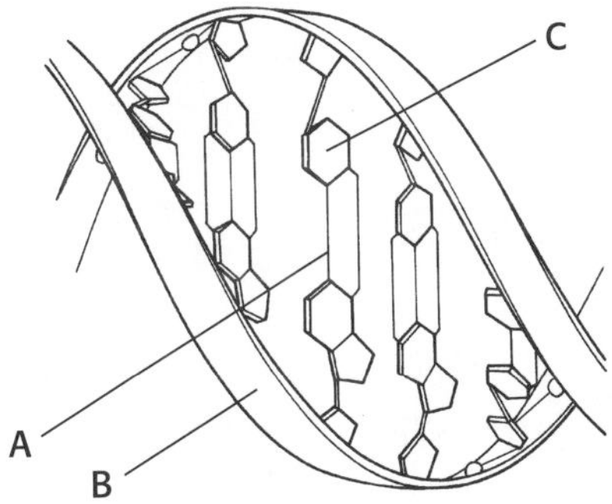

41. What does the figure above represent?

_____

_____

42. Identify the structures labeled *A–C*.

_____

_____

CHAPTER
9 **VOCABULARY**

DNA: The Genetic Material

Complete the crossword puzzle using the clues provided.

ACROSS

2. five-carbon sugar found in DNA nucleotides

4. enzyme that adds nucleotides to exposed nitrogen bases

5. substance prepared from killed or weakened microorganisms

6. change in phenotype of bacteria caused by the presence of foreign genetic material

8. The term "double _____" is used to describe the shape of DNA.

10. a virus that infects bacteria

11. enzyme that separates DNA by breaking the hydrogen bonds that link the nitrogen bases

12. name for DNA subunit

DOWN

1. relationship of two DNA strands to each other

3. Base-_____ rules describe the arrangement of the nitrogen bases between two DNA strands.

5. disease-causing

7. the process by which DNA is copied

9. A replication _____ is the area that results after the double helix separates during replication.

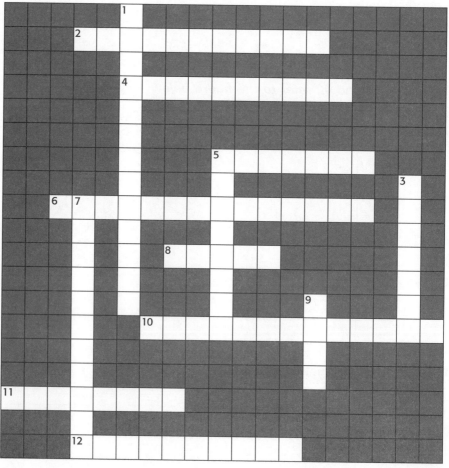

CHAPTER

9 **SCIENCE SKILLS: INTERPRETING DIAGRAMS**

DNA: The Genetic Material

Use the figure below to answer questions 1–4.

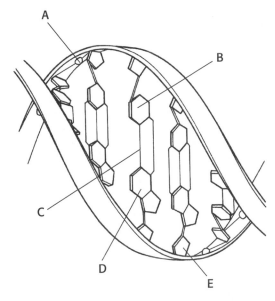

Read each question, and write your answer in the space provided.

1. In the space provided, identify the structures labeled *A–E*.

   A. _____

   B. _____

   C. _____

   D. _____

   E. _____

2. What do the lines connecting the two strands represent? Why are there three lines connecting the strands in some instances and only two lines in others?

   _____

   _____

   _____

   _____

3. Suppose that a strand of DNA has the base sequence ATT-CCG. What is the base sequence of the complementary strand?

   _____

4. Using the figure on page 67 as a model, draw a DNA model being separated at its replication fork and draw new nucleotides being added by DNA polymerase. Use circles to represent the nucleotides. Label the parts of your diagram, and color them according to the key below.

adenine—green
guanine—yellow
cytosine—purple

thymine—red
sugar—brown
phosphate group—blue

How Proteins Are Made

In the space provided, write the letter of the term or phrase that best completes each statement or best answers each question.

_____ 1. Which of the following is NOT a form of RNA?

    a. messenger RNA   **c.** transfer RNA

    b. ribosomal RNA   **d.** translation RNA

_____ 2. During transcription, the genetic information for making a protein is rewritten as a molecule of

    a. messenger RNA.   **c.** transfer RNA.

    b. ribosomal RNA.   **d.** translation RNA.

_____ 3. All organisms have a genetic code made of

    a. two-nucleotide sequences.   **c.** four-nucleotide sequences.

    b. three-nucleotide sequences.   **d.** five-nucleotide sequences.

_____ 4. In a cell, the equipment for translation is located in the

    a. cytoplasm.   **c.** plasma membrane.

    b. nucleus.   **d.** centrioles.

_____ 5. Gene regulation is necessary in living organisms

    a. so that the repressor will never bind to the operator.

    b. to allow RNA polymerase continuous access to genes.

    c. to avoid wasting their energy and resources on producing proteins that are not needed or are already available.

    d. to ensure that the operon is always in the "on" mode.

_____ 6. The *lac* operon enables a bacterium to build the proteins needed for lactose metabolism only when

    a. glucose is present.   **c.** galactose is present.

    b. tryptophan is present.   **d.** lactose is present.

_____ 7. Which of the following is NOT true about gene regulation in eukaryotic cells?

    a. Gene regulation in eukaryotes is more complex than in prokaryotes.

    b. Operons play a major role in eukaryote gene regulation.

    c. Gene regulation can occur before, during, or after transcription.

    d. Gene regulation can occur after translation.

_____ 8. Insertion or deletion mutations

    a. occur when one nucleotide is replaced with a different nucleotide.

    b. occur when, for example, the codon UGU is changed to UGC.

    c. upset the triplet nucleotide groupings within a gene.

    d. occur only in gametes.

•••••••••••••

Circle T *if the statement is true or* F *if it is false.*

T F **9.** Gene expression occurs in two phases: transcription and transformation.

T F **10.** In eukaryotic cells, the RNA nucleotides are found in the nucleus.

T F **11.** Transcription follows the same base-pairing rules as DNA replication without exception.

T F **12.** It appears that all life-forms had a common evolutionary ancestor with a single genetic code.

T F **13.** The anticodon of tRNA is so named because its three-nucleotide sequence is complementary to one of the 64 codons of the genetic code.

T F **14.** When lactose is present in bacterial cells, it prevents the blocking effect of the repressor protein and allows the transcription of lactose-metabolizing genes.

T F **15.** An enhancer is a sequence of nucleotides that, when activated by specific signal proteins, aids in shielding the RNA polymerase binding site of a specific gene.

T F **16.** Mutations in body cells can be passed on to the offspring of the affected individual.

T F **17.** An example of a single nucleotide mutation is seen in the genetic disorder sickle cell anemia.

T F **18.** If a point mutation changes the original amino acid sequence of the protein, the protein may not function normally.

•••••••••••••

Complete each statement by writing the correct term or phrase in the space provided.

19. Instead of the base thymine found in DNA, RNA has a base called

_____ .

20. Transcription begins when an enzyme called _____

_____ binds to the beginning of a gene on a region of DNA called a promoter.

21. The instructions for building a protein are written as a series of three-

nucleotide sequences called _____ .

22. During translation, the area of the ribosome called the _____ site receives the next tRNA molecule.

23. Because of its position on the operon, the _____ is able to control RNA polymerase's access to the structural genes.

24. The *lac* operon is switched off when a protein called a(n) _____ is bound to the operator.

DNA   RNA   Protein

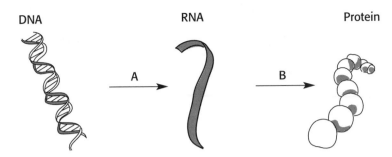

25. The processing of information from DNA into proteins, as shown above,

is referred to as _____ _____ .

26. Stage *A* is called _____ .

27. Stage *B* is called _____ .

28. In eukaryotic gene regulation, proteins called _____

_____ help arrange RNA polymerases in the correct position
on the promoter.

29. In eukaryotes, long segments of nucleotides with no coding information are

called _____ .

30. In eukaryotes, the portions of a gene that are actually translated into proteins

are called _____ .

31. Substitutions, insertions, and deletions are types of _____

_____ .

32. Because they cause the reading sequence to shift one or two positions, deletion

and insertion mutations are known as _____ mutations.

• • • • • • • • • • • • • • •
Read each question, and write your answer in the space provided.

33. Briefly explain how RNA differs from DNA.

_____

_____

_____

_____

34. Summarize the process of translation.

_____

_____

_____

35. Briefly describe the functions of RNA.

_____

_____

_____

_____

_____

36. What is the _lac_ operon?

_____

_____

_____

37. Explain why gene regulation in eukaryotic cells is more complex than in prokaryotic cells.

_____

_____

_____

_____

38. Why do scientists think that introns and exons contribute to evolutionary flexibility?

_____

_____

_____

_____

39. Briefly describe the two kinds of point mutations.

_____

_____

_____

_____

40. In what kinds of cells do mutations occur?

_____

_____

CHAPTER
(10) VOCABULARY

How Proteins Are Made

Complete the crossword puzzle using the clues provided.

ACROSS

2. The _____ operon controls the metabolism of lactose.

3. _____ acid (RNA) is a molecule made of nucleotides linked together.

4. RNA _____ is an enzyme involved in transcription.

5. _____ RNA carries the instructions for making a protein from a gene to the site of translation.

6. _____ expression is the entire process by which proteins are made.

8. process for transferring a gene's instructions for making a protein to an mRNA molecule

12. a three-nucleotide sequence on the mRNA that specifies an amino acid or "start" or "stop" signal

14. piece of DNA that serves as an on-off switch for transcription

16. long segment of nucleotides on a eukaryotic gene that has no coding information

DOWN

1. a nitrogen base in RNA that is complementary to thymine

3. a protein that binds to an operator and inhibits transcription

7. portion of a eukaryotic gene that is translated

8. a process that puts together the amino acids that make up a protein

9. a three-nucleotide sequence on a tRNA that is complementary to one of the codons of the genetic code

10. _____ RNA molecules are part of the structure of ribosomes.

11. _____ RNA molecules temporarily carry a specific amino acid on one end.

13. The _____ code specifies the amino acids and "start" and "stop" signals with their codon.

15. collective name for a group of genes involved in the same function, their promoter site, and their operator

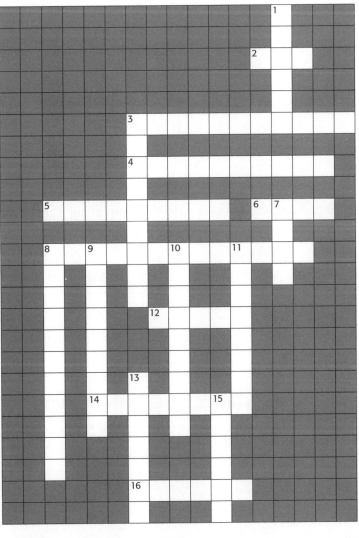

CHAPTER
10 **SCIENCE SKILLS: INTERPRETING TABLES**

How Proteins Are Made

Use the table below to complete items 1–3.

Codons in mRNA					
First base	**Second base**				**Third base**
	U	**C**	**A**	**G**	
U	UUU ⎤ Phenylalanine UUC ⎦ UUA ⎤ Leucine UUG ⎦	UCU ⎤ UCC ⎥ Serine UCA ⎥ UCG ⎦	UAU ⎤ Tyrosine UAC ⎦ UAA ⎤ Stop UAG ⎦	UGU ⎤ Cysteine UGC ⎦ UGA – Stop UGG – Tryptophan	**U** **C** **A** **G**
C	CUU ⎤ CUC ⎥ Leucine CUA ⎥ CUG ⎦	CCU ⎤ CCC ⎥ Proline CCA ⎥ CCG ⎦	CAU ⎤ Histidine CAC ⎦ CAA ⎤ Glutamine CAG ⎦	CGU ⎤ CGC ⎥ Arginine CGA ⎥ CGG ⎦	**U** **C** **A** **G**
A	AUU ⎤ Isoleucine AUC ⎥ AUA ⎦ AUG – Start	ACU ⎤ ACC ⎥ Threonine ACA ⎥ ACG ⎦	AAU ⎤ Asparagine AAC ⎦ AAA ⎤ Lysine AAG ⎦	AGU ⎤ Serine AGC ⎦ AGA ⎤ Arginine AGG ⎦	**U** **C** **A** **G**
G	GUU ⎤ GUC ⎥ Valine GUA ⎥ GUG ⎦	GCU ⎤ GCC ⎥ Alanine GCA ⎥ GCG ⎦	GAU ⎤ Aspartic acid GAC ⎦ GAA ⎤ Glutamic acid GAG ⎦	GGU ⎤ GGC ⎥ Glycine GGA ⎥ GGG ⎦	**U** **C** **A** **G**

1. Complete the table below showing sequences of DNA, mRNA codons, anti-codons, and corresponding amino acids. Use the list of mRNA codons in the table above to assist you in completing this exercise. Remember that the genetic code is based on mRNA codons.

Decoding DNA				
DNA	a. _____	b. _____	GAT	c. _____
mRNA codon	d. _____	e. _____	f. _____	UAU
Anticodon	g. _____	UUC	h. _____	i. _____
Amino acid	Tryptophan	j. _____	k. _____	l. _____

2. Determine how the mutations below will affect each amino acid sequence. Use the mRNA codons in the table on page 74 to complete items a–d below. In the space provided, write the names of the amino acids that correspond to each mRNA sequence and mutation given. An example is provided for you.

Example:

mRNA sequence:	UGU-CCG	cysteine-proline
mutation sequence:	UGC-CGC	cysteine-arginine

a. mRNA sequence: GAA-CGU   _____

   mutation sequence: GAU-CGU   _____

b. mRNA sequence: AUC-UGC   _____

   mutation sequence: AUC-UGG   _____

c. mRNA sequence: UGU-CCU-CCU   _____

   mutation sequence: UGU-UUC-CCU   _____

d. mRNA sequence: GGG-UUA-ACC   _____

   mutation sequence: GGU-UAA   _____

3. What kind of mutation occurred to the mRNA sequence in item *d* above? Explain.

_____

_____

_____

_____

_____

_____

Gene Technology

*In the space provided, write the letter of the term or phrase that
best completes each statement or best answers each question.*

_____ 1. Which of the following is NOT currently an application of genetic
engineering?
   a. production of interferon   **c.** use of a malaria vaccine
   b. production of anticoagulants   **d.** use of gene therapy

_____ 2. A powerful weapon in fighting cancer has been created by combining a
_____ with a kind of white blood cell that is very effective at locating
cancer cells.
   a. tumor necrosis factor gene   **c.** surface protein
   b. frog cell   **d.** Ti plasmid

_____ 3. Scientists can now genetically engineer vaccines by inserting the genes
that encode the pathogen's surface proteins into
   a. a specific pathogenic microbe.
   b. the DNA of harmless bacteria or viruses.
   c. the DNA of the patient.
   d. the RNA of the defective genes.

_____ 4. Radioactive or fluorescent-labeled RNA or single-stranded DNA pieces that
are complementary to the gene of interest and are used to confirm the
presence of a cloned gene are called
   a. probes.   **c.** vaccines.
   b. plasmids.   **d.** clones.

_____ 5. Which of the following is a potential benefit of genetically altered crops?
   a. Pesticide use is reduced.
   b. Tolerance of environmental stresses is increased.
   c. Food spoilage is reduced.
   d. all of the above

_____ 6. Which of the following is NOT an example of gene technology used in farming?
   a. the use of cow growth hormone produced by bacteria to increase
      milk production in cows
   b. the development of larger- and faster-growing breeds of livestock
   c. the cloning of human brain cells from selected farm animals
   d. the addition of human genes to farm-animal genes to obtain milk
      containing human proteins

_____ 7. Ian Wilmut's cloning of the sheep "Dolly" in 1997 was considered a
breakthrough in genetic engineering because
   a. scientists thought cloning was impossible.
   b. scientists thought only fetal cells could be used to produce clones.
   c. scientists had never before isolated mammary cells.
   d. sheep had never responded well to gene technology procedures.

Questions 8–10 refer to the figure below, which shows the four distinct steps of a genetic engineering experiment using DNA from a human insulin gene.

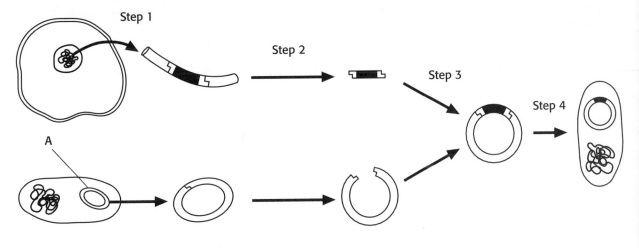

_____ 8. The structure labeled *A* is called
a. plasmid DNA.
b. a vector.
c. a restriction enzyme.
d. Both (a) and (b)

_____ 9. In Step 3, the DNA of the gene and the vector are
a. cloned.
b. isolated.
c. recombined.
d. cut by the restriction enzyme.

_____ 10. In Step 4, the
a. gene is cloned.
b. cells are screened.
c. recombined plasmid DNA is inserted into the bacterium.
d. DNA is cut.

• • • • • • • • • • • • • • •
In the space provided, write the letter of the description that best matches the term or phrase.

_____ 11. genetic engineering

_____ 12. vector

_____ 13. plasmid

_____ 14. cloning

_____ 15. recombinant DNA

_____ 16. restriction enzymes

_____ 17. DNA fingerprinting

_____ 18. Southern blot

_____ 19. gel electrophoresis

_____ 20. transgenic animal

_____ 21. differentiated cell

a. an animal with foreign DNA in its cells
b. made from DNA from two separate organisms
c. is programmed to become a specific type of cell
d. uses an electrical field to separate molecules
e. growing a large number of genetically identical cells from a single cell
f. isolates a gene from one organism's DNA and recombines it with another organism's DNA
g. pattern of dark bands on X-ray film made when DNA fragments are separated and probed
h. uses radioactively labeled RNA or single-stranded DNA as a probe to identify specific genes
i. can carry a DNA fragment into another cell
j. can be used as a vector
k. used to cut DNA at specific sequences

Complete each statement by writing the correct term or phrase in the space provided.

22. The first step of Cohen and Boyer's genetic engineering experiment was to isolate

the _____ of interest from the DNA of an African clawed frog.

23. Recombinant DNA is made when a DNA fragment is put into the DNA of a(n)

_____ .

24. Any two fragments of DNA cut by the same restriction enzyme can pair because

their ends are _____ .

25. Genetic engineering has benefited humans afflicted with diabetes by developing

bacteria that produce _____ .

26. _____ produced by genetic engineering can now be used to help dissolve blood clots in people who have suffered heart attacks.

27. By using the genetically engineered blood-clotting agent _____

_____ , hemophiliacs can eliminate the risks associated with blood products obtained from other individuals.

28. A vaccine is a solution that contains a weakened or modified version of a(n)

_____ or its toxins.

29. Genetic engineering turned a tumor-causing _____

_____ into a suitable vector for broad-leaved crop plants.

30. Crop plants that are resistant to the biodegradable herbicide

_____ have been developed.

31. In genetic engineering, the enzyme _____

_____ helps the DNA fragments bind.

Read each question, and write your answer in the space provided.

32. Briefly describe the Human Genome Project.

_____

_____

_____

_____

33. List two ways in which DNA fingerprints are used.

_____

_____

_____

34. Explain why the development of genetically engineered proteins has been important to pharmaceutical companies.

_____

_____

_____

_____

_____

_____

35. How did scientists develop the tumor-causing Ti plasmid into a genetic engineering vehicle for plants?

_____

_____

_____

_____

_____

_____

36. Why is the development of plants that are resistant to insects especially important?

_____

_____

_____

37. Explain the first step of Ian Wilmut's successful cloning experiment.

_____

_____

_____

_____

_____

Name_____ Date _____ Class _____

Gene Technology

Unscramble each listed term, and write the correct term in the space at right. In the space at left, write the letter of the description below that best matches the term or phrase.

_____ **1.** nmauH nmeeGo tjPceor _____

_____ **2.** tveorc _____

_____ **3.** smipdal _____

_____ **4.** brtnneacimo DAN _____

_____ **5.** ccveina _____

_____ **6.** ttiicrneosr meenyzs _____

_____ **7.** neeg atyhrep _____

_____ **8.** neeg gninolc _____

_____ **9.** ssierhpocrteelo _____

_____ **10.** raeicngsnt aaimnl _____

_____ **11.** eegncit ggeeiinnren _____

a. animal that has foreign DNA in its cells

b. a solution containing a weakened or modified version of a pathogen

c. a research effort to determine the nucleotide sequence of the human genome and map the location of every gene

d. a technique that involves putting a healthy copy of a gene into the cells of a person whose copy of the gene is defective

e. bacterial enzymes that recognize and bind to specific short sequences of DNA, then cut the DNA at specific sites within the sequences

f. a technique that uses an electrical field within a gel to separate molecules by their size and charge

g. a circular DNA molecule that can replicate independently of the main chromosomes of bacteria

h. an agent that is used to carry the gene of interest into another cell

i. DNA made from two or more different organisms

j. when copies of the gene of interest are made each time the host cell reproduces

k. the process of manipulating genes for practical purposes

Name_____ Date_____ Class_____

Gene Technology

Most genetic engineering experiments include four basic steps, as shown in the figure below. In this example, the gene responsible for producing insulin is the gene of interest. Use the figure below to answer questions 1–4 below.

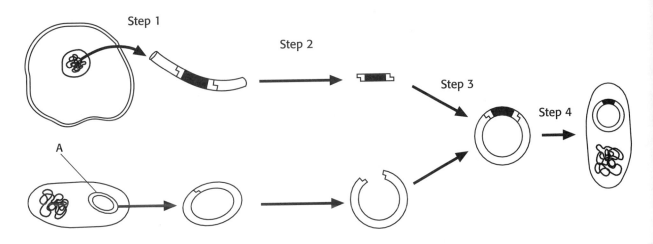

Read each question, and write your answer in the space provided.

1. Identify the structure labeled *A*. What is the function of this structure in Step 1?

_____

_____

2. Explain Step 2 of the figure above.

_____

_____

_____

3. Explain Step 3 of the figure above.

_____

_____

_____

4. What is the final product in this process, as shown in Step 4?

_____

_____

Restriction enzymes recognize specific short nucleotide sequences in DNA and cut them within those sequences, resulting in single stranded areas called sticky ends, as shown in the figure below. In order for the plasmid DNA to recombine successfully with the gene of interest, the sticky ends of the plasmid DNA must pair with the complementary sticky ends of the gene of interest. Use the figure below to answer questions 5–7 below.

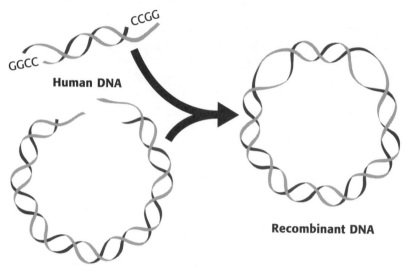

Human DNA

Plasmid DNA

Recombinant DNA

Read each question, and write your answer in the space provided.

5. List the two nucleotide sequences that are complementary to the sticky end sequences on the human DNA.

_____

6. List the paired nucleotide sequences of the recombinant DNA.

_____

7. The gene for tetracycline resistance is present in the plasmid DNA. Explain the reason for using plasmid DNA that contains the gene for tetracycline resistance.

_____

_____

_____

_____

_____

_____

_____

_____

CHAPTER

12 **TEST PREP PRETEST**

History of Life on Earth

In the space provided, write the letter of the term or phrase that best completes each statement or best answers each question.

_____ 1. Radiometric dating has determined that Earth is approximately
    a. 4,000 years old.
    b. 500,000 years old.
    c. 2.5 billion years old.
    d. 4.5 billion years old.

_____ 2. Primitive cyanobacteria
    a. were among the first bacteria to appear on Earth.
    b. produced the first oxygen in Earth's atmosphere.
    c. are believed to be the ancestors of chloroplasts.
    d. All of the above

_____ 3. A mechanism for heredity was necessary in order to begin
    a. microspheres.
    b. life.
    c. RNA.
    d. protein.

_____ 4. The kingdoms that evolved from protists are
    a. bacteria, fungi, and animals.
    b. bacteria, plants, and animals.
    c. plants, animals, and humans.
    d. fungi, plants, and animals.

_____ 5. Life was able to move from the sea to land because
    a. photosynthesis by cyanobacteria added oxygen to Earth's atmosphere.
    b. ozone was created from the oxygen produced by photosynthesis.
    c. ozone provides a shield from the harsh ultraviolet rays of the sun.
    d. All of the above

_____ 6. Radiometric dating compares the proportions of specific radioisotopes with their
    a. atomic mass.
    b. more stable isotopes.
    c. charged particles.
    d. unstable isotopes.

_____ 7. In a unique biological association known as mycorrhizae, fungi provide
    a. oxygen to plants through photosynthesis.
    b. food to plants and the plants provide minerals to the fungi.
    c. minerals to plants and the plants provide food to the fungi.
    d. All of the above

_____ 8. Louis Lerman's bubble model addresses the problem of
    a. lightning.
    b. ultraviolet radiation damage to ammonia and methane.
    c. spontaneous origin.
    d. a dense ozone layer.

_____ **9.** The following animals are all arthropods.

    a. crabs, lobsters, insects, and spiders
    b. crabs, snakes, insects, and spiders
    c. frogs, toads, and salamanders
    d. frogs, toads, salamanders, and snakes

_____ **10.** The first vertebrates on land were

    a. reptiles.   **c.** amphibians.
    b. insects.   **d.** fishes.

_____ **11.** The primordial soup model of spontaneous origin requires

    a. the sun, electrical energy, or volcanic eruptions.
    b. at least one billion years.
    c. hydrogen-containing gases.
    d. All of the above

_____ **12.** Amphibians were able to adapt to life on land for all of the following reasons EXCEPT

    a. lungs.
    b. watertight skin.
    c. limbs.
    d. a platform of bone that provided a base for the limbs to work against.

_____ **13.** Reptiles are more completely adapted to land than amphibians are because reptiles

    a. have watertight skin.
    b. can lay their eggs on dry land.
    c. do not have to live near the water.
    d. All of the above

_____ **14.** Scientists think the first step toward cellular organization was

    a. nucleotides.   **c.** microspheres.
    b. coacervates.   **d.** RNA enzymes.

• • • • • • • • • • • • • •

Circle T *if the statement is true or* F *if it is false.*

T F **15.** Bacteria are thought to have evolved 3.5 billion years ago.

T F **16.** Early cyanobacteria were responsible for much of the oxygenation of the oceans.

T F **17.** Early cyanobacteria are thought to be ancestors of chloroplasts.

T F **18.** Mass extinctions cause increased competition for resources among the survivors.

T F **19.** *Escherichia coli*, a species of eubacteria, has peptidoglycan in its cell walls.

T F **20.** In the upper atmosphere, ozone blocks the sun's harsh ultraviolet radiation, preventing it from reaching the surface of Earth.

T F **21.** The first multicellular creatures to live on land were plants and bacteria.

T F **22.** Mycorrhizae are associations between fungi and the roots of plants.

T F **23.** The lack of a protective ozone layer weakens the primordial soup model.

T F **24.** RNA, not DNA, may have been the first self-replicating information storage molecule.

T F **25.** The first vertebrates were small, jawless fishes.

T F **26.** The theory of endosymbiosis proposes that mitochondria arose from aerobic archaebacteria.

T F **27.** Multicellularity, which enabled cell specialization, developed first in protists.

T F **28.** Reptiles evolved from amphibians.

T F **29.** Continental drift explains why related animals are found on continents separated by oceans.

T F **30.** Birds and mammals have become the dominant vertebrates on land partly because of a shift toward a moister climate on Earth.

• • • • • • • • • • • • • •

Complete each statement by writing the correct term or phrase in the space provided.

31. The earliest traces of life on Earth are fossils of _____ .

32. _____ appear to be the ancestors of all eukaryotic cells.

33. In Louis Lerman's model of spontaneous origin, the hydrogen-containing gases needed to make amino acids were trapped in _____

_____ .

34. The earliest stages of cellular organization may have been the formation of

_____ .

35. Because of the mass extinction at the end of the Permian period, about

96 percent of all species of _____ living at the time became extinct.

36. The formation of _____ from oxygen in the upper atmosphere made Earth's surface a safe place to live for the first time.

37. The first multicellular organisms to live on land were _____

and _____ .

38. A partnership, such as the one found in mycorrhizae, in which each organism

helps the other survive is called _____ .

39. The first group of animals to live on land was the _____ .

40. Among the earliest _____ were the _____ , which released oxygen.

41. Evidence that eukaryotes evolved from bacterial invasions of pre-eukaryotic

cells includes similarities of size and structure; _____

_____ ; ribosomes; and methods of _____

between bacteria, mitochondria, and chloroplasts.

42. How was the oxygen in Earth's atmosphere generated?

_____

_____

_____

_____

43. Why was the development of multicellular organisms a great step forward in the evolution of life on Earth?

_____

_____

44. What were the first vertebrates to live on land? What structural changes enabled them to make the transition?

_____

_____

_____

_____

_____

45. Briefly explain the problems with the primordial soup model.

_____

_____

_____

_____

46. List in order, from first to most recent, the four groups of organisms to invade land.

_____

_____

47. If snakes and frogs both lived in the same drought-stricken area, which species would be most likely to survive? Give at least two reasons.

_____

_____

_____

Name ______________________ Date ____________ Class ____________

CHAPTER 12 VOCABULARY

History of Life on Earth

In the space provided, explain how the terms in each pair differ in meaning.

1. radioisotope, half-life

2. spontaneous origin, microsphere

3. eubacteria, archaebacteria

4. mycorrhiza, mutualism

(continued on next page)

In the space provided, write the letter of the description that best matches the term or phrase.

_____ 5. radiometric dating

_____ 6. fossil

_____ 7. cyanobacteria

_____ 8. endosymbiosis

_____ 9. protists

_____ 10. mass extinction

_____ 11. arthropod

_____ 12. vertebrate

_____ 13. continental drift

a. animal with hard outer skeleton and jointed limbs

b. the movement of Earth's land masses over geologic time

c. animals with backbones

d. calculation of the age of an object by measuring the proportions of radioactive isotopes of certain elements

e. the preserved or mineralized remains or imprint of an organism that lived long ago

f. among the first bacteria to appear

g. the theory that mitochondria are the descendants of symbiotic aerobic eubacteria

h. members of a kingdom of unicellular and multicellular eukaryotic organisms

i. the death of all members of many different species

SCIENCE SKILLS: INTERPRETING TIMELINES

History of Life on Earth

The timeline below shows some of the physical events that have helped to shape life on Earth. Some major advances in the evolution of life are also shown. Use the timeline below and the "Life Events" (a–d) listed below to complete items 1–4.

In each numbered blank in the timeline, write the letter of the most appropriate event from the "Life Events" below.

Timeline of Life on Earth

Number of years ago	Life events	Physical events
4.5 billion		Earth forms and surface cools.
3.5 billion	The first bacteria evolve in the sea.	
3 billion	1. _____	Oxygen gas is released into seas, then enters atmosphere.
1.5 billion	The first eukaryotes evolve.	
700 million	The first multicellular organisms evolve.	
440 million	First mass extinction	
430 million	2. _____	Protective ozone shield in place in upper atmosphere
370 million	Amphibians are the first vertebrates on land.	
360 million	Second mass extinction	
350 million	3. _____	Widespread drought
245 million	Third mass extinction	
210 million	Fourth mass extinction	
65 million	Fifth mass extinction	
60 million	4. _____	Earth's climate becomes moist.
2 million	The first humans appear.	
Present	More than half of tropical rain forests and many species destroyed	

Life Events

a. Reptiles evolve from amphibians and become dominant on land.

b. Birds and mammals become dominant on land.

c. Plants and fungi invade land for the first time.

d. Cyanobacteria first begin to carry out photosynthesis.

Use the timeline on page 89 to answer questions 5–9.

Read each question, and write your answer in the space provided.

5. Name the only organism that has seriously damaged Earth's ecosystems, and describe the damage caused by this organism.

_____

_____

_____

6. What factors have caused the dominant life-forms on Earth to change over time?

_____

_____

_____

_____

7. Is another major mass extinction likely? Explain.

_____

_____

8. What environmental changes affected the evolution of reptiles?

_____

_____

9. What effect did photosynthetic cyanobacteria have on Earth's environment?

_____

_____

_____

CHAPTER

13 **TEST PREP PRETEST**

The Theory of Evolution

In the space provided, write the letter of the term or phrase that best completes each statement or best answers each question.

_____ 1. The part of Lamarck's hypothesis of evolution that proved to be correct was that
   a. evolution is linked to an organism's environmental conditions.
   b. evolution relies on the use and disuse of physical features.
   c. acquired traits are passed on to offspring.
   d. evolution is a slow process of gradual change.

_____ 2. On the Galápagos Islands, Darwin saw that the plants and animals closely resembled those of the
   a. islands off the coast of North America.
   b. coast of South America.
   c. islands off the coast of Africa.
   d. coast of South Africa.

_____ 3. Darwin developed the idea of natural selection by
   a. modifying traditional accounts of creation.
   b. taking Lamarck's hypothesis on evolution as his own.
   c. studying for the ministry at Cambridge University.
   d. applying Malthus's ideas on population to the observations he made during the voyage of the *Beagle*.

_____ 4. Which of the following is a factor in natural selection?
   a. Individuals of a species compete with one another to survive.
   b. All species are genetically diverse.
   c. Individuals better able to adapt to changes leave more offspring.
   d. all of the above

_____ 5. When the individuals of two populations can no longer interbreed, the two populations are considered to be
   a. different families.   **c.** the same species.
   b. different species.   **d.** unrelated.

_____ 6. The fossil record provides evidence that
   a. older species gave rise to more-recent species.
   b. all species were formed during Earth's formation and have changed little since then.
   c. the fossilized species have no connection to today's species.
   d. fossils cannot be dated accurately.

_____ 7. Comparing human hemoglobin with the hemoglobin of gorillas, mice, chickens, and frogs reveals that humans have the fewest amino acid differences with
   a. gorillas.   **c.** chickens.
   b. mice.   **d.** frogs.

8. Individuals that are better able to cope with the challenges of their environment tend to

   a. decrease in population over time.
   b. leave more offspring than those more suited to the environment.
   c. leave fewer offspring than those less suited to the environment.
   d. leave more offspring than those less suited to the environment.

9. In response to the darkening of tree trunks by pollution in England, some European peppered moth populations evolved from

   a. dark gray to cream-colored.
   b. dark gray to yellow.
   c. cream-colored to yellow.
   d. cream-colored to dark gray.

10. Which factor does NOT play a role in determining the beak size of Galápagos finches?

   a. amount of food available   **c.** size of the bird
   b. seed size   **d.** weather

11. Members of different ecological races

   a. are considered to be different species.
   b. differ genetically because of adaptations for different living conditions.
   c. can no longer interbreed successfully.
   d. will never diverge to become different species.

Questions 12 and 13 refer to the figures below.

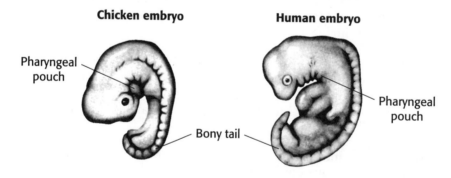

12. Which of the following statements best reflects the evolutionary importance of the figures above?

   a. New genetic instructions have been disregarded in the evolution of vertebrates.
   b. Early in development, vertebrate embryos show no evidence of common ancestry.
   c. The evolutionary history of organisms is evident in the way embryos develop.
   d. All adult vertebrates retain pharyngeal pouches.

13. Which of the following statements is NOT true about the vertebrate embryos shown above?

   a. Each embryo develops a tail.
   b. Each embryo has buds that become limbs.
   c. Each embryo has pharyngeal pouches.
   d. Each embryo has fur.

T F **14.** There is clear evidence from fossils and other sources that the species now on Earth have evolved from ancestral species that are extinct.

T F **15.** In South America, Darwin found fossils of armadillos that were identical to the armadillos that were living there.

T F **16.** Darwin combined Malthus's principles of population, the observations he made during the voyage of the *Beagle*, and his experience in breeding domestic animals to form the theory of evolution by natural selection.

T F **17.** Ecological races are genetically different but can interbreed.

T F **18.** Fossils have been found that provide a link in the evolution of whales from four-legged land animals.

T F **19.** Even though species may have changed over time, the genes that determine what they are have not changed.

T F **20.** The forelimbs of all vertebrates contain the same bones and serve the same function.

T F **21.** The fossil record seems to provide evidence for both gradualism and punctuated equilibrium.

T F **22.** The characteristics of the individuals best suited to a particular environment tend to increase in a population over time.

T F **23.** Kettlewell concluded that natural selection caused industrial melanism in peppered moth populations.

T F **24.** Divergence is the final step of speciation.

T F **25.** One mechanism for reproductive isolation is geographic isolation.

Complete each statement by writing the correct term or phrase in the space provided.

26. Over time, change within species leads to the replacement of old species by new

species as less successful species become _____ .

27. While on the *Beagle*, Darwin read *Principles of Geology*, which contained a

detailed account of _____ theory of evolution.

28. The changing of a species that results in its being better suited to its

environment is called _____ .

29. A(n) _____ is a group of individuals that belong to the same species, live in a defined area, and breed with others in the group.

30. The condition in which two populations of the same species are separated from

one another is called _____ .

31. Species that shared a common ancestor in the recent past have many

_____ _____ or _____

sequence similarities.

32. Given that the forelimbs of all vertebrates share the same basic arrangement of

bones, forelimbs are said to be _____ structures.

33. The _____ of individuals who adapt to changing conditions
tend to increase over time.

34. Paleontologists use _____ _____ to determine
the proper sequence of fossils from oldest to youngest.

35. The model of evolution in which gradual change leads to species formation over

time is called _____ .

36. A whale's pelvic bones are _____ structures because they no
longer function like the pelvis of a land vertebrate.

37. Darwin felt that fossils of extinct armadillos that resembled living armadillos

were evidence that _____ is a(n) _____
process.

38. _____ is the accumulation of differences between groups such
as populations, species, and genera.

• • • • • • • • • • • • • •
Read each question, and write your answer in the space provided.

39. What was Lamarck's hypothesis regarding evolution?

_____

_____

_____

40. Briefly explain the importance of Thomas Malthus's essay on the growth of the
human population to Darwin's theory of evolution.

_____

_____

_____

_____

41. Briefly summarize the modern version of Darwin's theory of evolution by
natural selection.

_____

_____

_____

_____

_____

CHAPTER

13 VOCABULARY

The Theory of Evolution

In the space provided, write the letter of the term or phrase that best completes each statement or best answers each question.

_____ 1. The process in which organisms with traits well suited to an environment are more likely to survive and to produce offspring is

  a. trait mechanisms.　　**c.** genetic principles.
  b. origin of species.　　**d.** natural selection.

_____ 2. In biology, all of the individuals of a species that live together in one place at one time are called a

  a. population.　　**c.** half-life.
  b. community.　　**d.** habitat.

_____ 3. A change in the genetic makeup of species over time is called

  a. radioactive dating.　　**c.** camouflage.
  b. evolution.　　**d.** natural selection.

_____ 4. The process by which a species becomes better suited to its environment is

  a. industrialization.　　**c.** adaptation.
  b. not an advantage.　　**d.** destructive to its survival.

_____ 5. Structures that share a common ancestry or are similar because they are modified versions of structures from a common ancestor are

  a. not related.　　**c.** not homologous.
  b. homologous.　　**d.** young in origin.

_____ 6. Structures with no function that are remnants of an organism's evolutionary past are

  a. not visible on organisms.　　**c.** vestigial.
  b. young in origin.　　**d.** useful to the organism.

_____ 7. The accumulation of differences between species or populations is called

  a. gradualism.　　**c.** divergence.
  b. punctuated equilibrium.　　**d.** observational species.

_____ 8. The hypothesis that evolution of a species occurs in periods of rapid change separated by periods of little or no change is called

  a. divergence.　　**c.** isolation.
  b. gradualism.　　**d.** punctuated equilibrium.

_____ 9. Populations of the same species that differ genetically because they have adapted to different living conditions are

  a. observational species.　　**c.** ecological races.
  b. different species.　　**d.** conditional races.

_____ 10. The hypothesis that the evolution of different species occurs at a slow constant rate is called

  a. punctuated equilibrium.　　**c.** divergence.
  b. gradualism.　　**d.** transitionism.

_____ 11. The condition in which two populations of the same species cannot breed with one another is called
    a. infertility.
    b. extinction.
    c. isolation.
    d. selection.

_____ 12. When a species permanently disappears, the species is said to be
    a. extinct.
    b. isolated.
    c. mutated.
    d. eliminated.

_____ 13. The darkening of populations over time in response to industrial pollution is called
    a. natural selection.
    b. evolution.
    c. industrial melanism.
    d. adaptation.

_____ 14. The process by which new species form is called
    a. biological change.
    b. reproduction.
    c. speciation.
    d. divergence.

_____ 15. The inability of formerly interbreeding groups to mate or produce fertile offspring is called
    a. sterility.
    b. divergence.
    c. reproductive isolation.
    d. extinction.

_____ 16. A scientist who studies fossils is called a(n)
    a. archaeologist.
    b. ecologist.
    c. paleontologist.
    d. biologist.

_____ 17. In Grant's study, the effect of weather on the size of the finch's beak is an example of
    a. isolation.
    b. natural selection.
    c. industrial melanism.
    d. fossilization.

_____ 18. Biological molecules that are considered evidence for evolution include
    a. DNA.
    b. amino acids.
    c. proteins.
    d. All of the above

Name _____ Date _____ Class _____

The Theory of Evolution

Darwin stated that evolution occurs through natural selection. The key factor is the environment. The environment "selects" which organisms will survive and reproduce. Traits possessed by organisms successful at survival and reproduction are more likely to be transmitted to the next generation. These traits will, therefore, become common.

Read the following information about the elephant population of Queen Elizabeth National Park in Uganda, Africa. Then use the table on page 98 to answer questions 1–5.

Normally, nearly all African elephants, male and female, have tusks. In 1930, only 1 percent of the elephant population in Queen Elizabeth National Park was tuskless because of a rare genetic mutation. Food was fairly plentiful, and by 1963, there were 3,500 elephants in the park. In the 1970s, a civil war began in Uganda. Much of the wildlife was killed for food, and poachers killed elephants for their ivory tusks. By 1992, the elephant population had dropped to about 200. But by 1998, the population had increased to 1,200. A survey in 1998 revealed that as many as 30 percent of the adult elephants did not have tusks. Ugandan wildlife officials also noted a decline in poaching.

(continued on next page)

In the space provided in the table below, explain how each of the given principles of natural selection applies to the situation described on page 97.

The Process of Natural Selection

Principles	Applications
All species have genetic variation.	1. _____ _____ _____
Living things face many challenges in the struggle to exist.	2. _____ _____ _____
Individuals of species often compete with one another to survive.	3. _____ _____ _____
Individuals that are better able to cope with the challenges of their environment tend to leave more offspring than those less suited.	4. _____ _____ _____
The characteristics of the individuals best suited to a particular environment tend to increase in a population over time.	5. _____ _____ _____

Name_____ Date _____ Class _____

Human Evolution

In the space provided, write the letter of the term or phrase that best completes each statement or best answers each question.

_____ 1. The two distinct anatomical changes that set the earliest primates apart from their ancestors were
   a. opposable thumbs and color vision.
   b. big eyes and clawed, unbendable toes.
   c. grasping fingers and binocular vision.
   d. color vision and grasping fingers.

_____ 2. When compared with monkeys, apes have
   a. larger, more developed brains.
   b. smaller, less developed brains.
   c. smaller, more developed brains.
   d. brains of a similar size.

Questions 3 and 4 refer to the figure below.

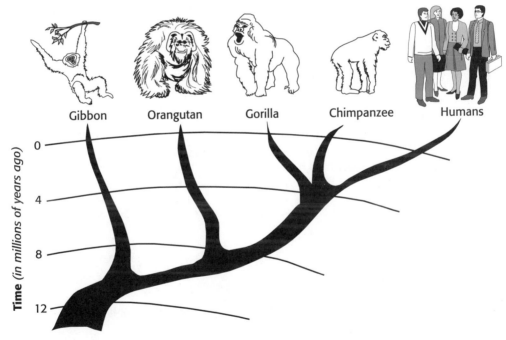

_____ 3. The evolutionary tree shown above indicates that the closest relative of humans is the
   a. chimpanzee.   **c.** orangutan.
   b. gorilla.   **d.** gibbon.

_____ 4. The split between gorillas and chimpanzees and the line leading to humans occurred about
   a. 4 million years ago.   **c.** 8 million years ago.
   b. 6 million years ago.   **d.** 10 million years ago.

_____ 5. Compared with modern humans, Neanderthals
    a. had a larger brain.
    b. had a more compact body.
    c. had heavy, bony ridges over their brow.
    d. All of the above

_____ 6. The ancestor of primates
    a. was a prosimian.
    b. evolved 50 million years ago.
    c. ate insects, had large eyes, and had small, sharp teeth.
    d. ate insects, had large forward-placed eyes, and had sharp teeth.

_____ 7. Australopithecines differed from apes in that they had
    a. arms longer than their legs and used both sets of appendages for walking.
    b. a skeletal structure that enabled them to walk upright.
    c. a tall, narrow pelvis.
    d. a skull that sat atop a C-shaped spine.

_____ 8. Fossil evidence that human ancestors walked upright includes
    a. the spinal cord emerging from the bottom of the skull.
    b. femurs placed directly below the body.
    c. an S-shaped spine.
    d. All of the above

_____ 9. Which of the following is NOT true about _Homo habilis_?
    a. evolved from an australopithecine ancestor in Africa about 2 million years ago
    b. had a much larger brain than australopithecines
    c. lived in Africa for 500,000 years and then migrated to Europe and Asia
    d. made and used stone tools

_____ 10. _Homo erectus_
    a. walked upright, used tools, and had a 1,000 cm^3 brain.
    b. is 1 million years older than Java man.
    c. gave rise to _Homo habilis_ and used tools.
    d. had a 640 cm^3 brain and walked upright.

_____ 11. The role of humans in the animal kingdom has been greatly affected by
    a. symbolic language.
    b. construction and use of tools.
    c. spoken language.
    d. All of the above

_____ 12. About 34,000 years ago, European Neanderthals were replaced by
    a. modern _Homo sapiens_.
    b. _Homo erectus_.
    c. _Homo habilis_.
    d. Peking man.

_____ 13. Raymond Dart's discovery, _Australopithecus africanus_, was considered
    a. an evolutionary link between hominids and primates.
    b. an early prehominid.
    c. an evolutionary link between apes and humans.
    d. the oldest _Australopithecus_.

In the space provided, write the letter of the description that best matches the term or phrase.

_____ 14. *Homo habilis*

_____ 15. prosimian

_____ 16. *Homo sapiens*

_____ 17. primate

_____ 18. diurnal

_____ 19. *Australopithecus afarensis*

_____ 20. binocular vision

_____ 21. *Australopithecus africanus*

_____ 22. Neanderthals

_____ 23. anthropoids

a. first hominids to show evidence of abstract thinking, caring for injured and sick, and burying their dead

b. active only during daylight hours

c. first hominid fossil ever found; showed evidence of bipedalism

d. preceded *A. africanus* by at least 1 million years; "Lucy" is most famous representative

e. member of a group of mostly nocturnal primates that includes lemurs, tarsiers, and lorises

f. enables depth perception

g. only living member of its genus

h. brain volume of about 640 cm^3; 1.2 m tall; associated with tools

i. member of a group that includes humans, apes, monkeys, and prosimians

j. a group that includes monkeys, apes, and humans

Complete each statement by writing the correct term or phrase in the space provided.

24. Natural selection favored an increase in the numbers of _____ cells, which made color vision and sharpened daytime vision possible.

25. The oldest hominid yet discovered belongs to the genus _____ and was found in Ethiopia.

26. About 36 million years ago, primates became _____ ; that is, they started feeding during the day and sleeping at night.

27. _____ were among the first primates with fully developed opposable thumbs.

28. *Homo* _____ became extinct after 500,000 years and had

a(n) _____ brain than the australopithecines.

29. Australopithecines, along with humans, are classified as _____ .

30. The _____ of australopithecines had greater volume relative to body weight than those of apes.

31. The first members of the genus _____ evolved from australopithecine ancestors about 2 million years ago.

32. The first hominids thought to have an advanced language are

_____ _____ .

33. The evolution of _____ , the ability to walk upright on two legs, was likely a response to environmental changes 15 million years ago.

••••••••••••••
Read each question, and write your answer in the space provided.

34. What evolutionary adaptations were evident in the first monkeys?

_____

_____

35. What two characteristics distinguished australopithecines from apes?

_____

_____

36. Briefly explain the two hypotheses of the origin of *Homo sapiens*.

_____

_____

_____

_____

_____

_____

37. When and how did *Homo sapiens* reach North America?

_____

_____

38. What evidence confirmed that Richard Leakey's 1972 fossil discovery was an early human and not an australopithecine?

_____

_____

39. Briefly explain how opposable thumbs affected the evolution of primates.

_____

_____

_____

CHAPTER

14 **VOCABULARY**

Human Evolution

In the space provided, write the letter of the description that
best matches the term or phrase.

_____ 1. primates

_____ 2. prosimians

_____ 3. diurnal

_____ 4. anthropoids

_____ 5. opposable thumb

_____ 6. hominids

_____ 7. bipedal

a. can be bent inward toward fingers to hold an object

b. active during the day and sleep at night

c. able to walk upright on two legs

d. group that includes prosimians, monkeys, apes, and humans

e. mostly night-active primates that live in trees

f. monkeys, apes, and humans

g. primates that can walk upright on two legs

Complete each statement by writing the correct term or phrase
in the space provided.

8. Monkeys are day-active _____ with opposable thumbs.

9. Australopithecines, unlike apes, were _____ and had large brains.

10. The first primates were _____ .

11. _____ primates are active during the day and sleep at night.

12. A(n) _____ _____ gives a hand a greatly increased level of skill at manipulating objects.

13. The members of the group _____ led to the evolution of humans.

Human Evolution

There are two major hypotheses concerning the success of modern humans, *Homo sapiens*, over the early hominid *Homo erectus*. Fossils of *H. erectus* have been found in Africa, Asia, and Europe, as shown in the map below. Some scientists think that several isolated populations of *H. erectus* gave rise to *H. sapiens* independently of each other around the same time period. Other scientists argue that *H. sapiens* arose in one place and that *H. erectus* populations across the world were replaced by this newer, superior species, which now inhabits all regions of the Earth.

It is generally agreed that *H. sapiens* was the first hominid inhabitant of North America and that the species crossed the land bridge that once existed between Siberia and Alaska at least 12,000 years ago.

Study the map below, which illustrates sites in Africa, Asia, and Europe where fossils of *Homo erectus* were found.

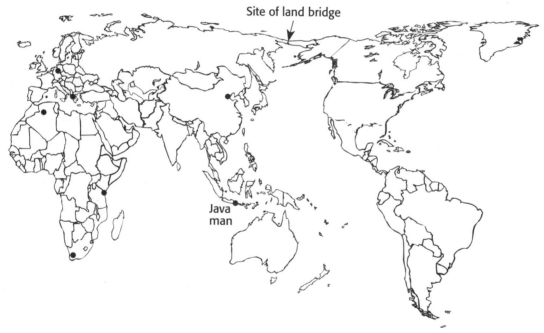

Site of land bridge

Java man

• Site of *Homo erectus* fossils

Bearing these data in mind, suppose the following important discoveries have recently been made. Use these findings to answer question 1 on page 105.

   A. An *H. erectus* skull thought to be 1 million years old is found in Canada.

   B. A Neanderthal skull thought to be 100,000 years old is found in Arizona.

   C. New geological techniques reveal that the land bridge from Siberia to Alaska disappeared at least 750,000 years ago.

Read each question, and write your answer in the space provided.

1. Which hypothesis about the evolution of *H. sapiens* would these findings support? Explain.

_____

_____

_____

_____

_____

_____

_____

Now suppose that a flurry of excavations has produced a very complete fossil record, which leads to the following findings. Use these findings to answer question 2 below.

A. The australopithecines (prehuman hominids) were apparently confined to Africa.

B. *Homo habilis*, the earliest member of the genus *Homo*, was also apparently confined to Africa.

C. *H. erectus* was found in Africa, Asia, and Europe, as is illustrated in the map on page 104.

D. Nothing predating *H. sapiens* has been found in North or South America.

2. Which hypothesis about the evolution of *H. sapiens* would these findings support? Explain.

_____

_____

_____

_____

_____

_____

_____

CHAPTER

15 **TEST PREP PRETEST**

Classification of Organisms

In the space provided, write the letter of the term or phrase that best completes each statement or best answers each question.

_____ 1. In one of the earliest classification systems, Aristotle grouped plants and animals according to

    a. basic categories.
    b. structural similarities.
    c. genus.
    d. major characteristics.

_____ 2. Although Linnaeus used the Latin polynomial system in his books, he created his own

    a. rules of grammar.
    b. taxonomic categories.
    c. evolutionary systematics.
    d. two-word shorthand system, also in Latin.

_____ 3. Scientists classify organisms by studying their forms and

    a. structures.
    b. size.
    c. method of reproduction.
    d. cladograms.

_____ 4. Cladograms determine evolutionary relationships between organisms by examining similar

    a. behavioral traits.
    b. morphological characteristics.
    c. molecular properties.
    d. All of the above

_____ 5. The basic biological unit in Linnaeus's system is the

    a. genus.
    b. family.
    c. species.
    d. class.

_____ 6. All members of the kingdom Animalia are multicellular

    a. autotrophs whose cells have walls.
    b. heterotrophs whose cells have walls.
    c. heterotrophs whose cells lack walls.
    d. autotrophs whose cells lack walls.

_____ 7. Biological species, as defined by Ernst Mayr,

    a. are closely related.
    b. are interbreeding natural populations.
    c. produce infertile offspring.
    d. produce infertile hybrids.

_____ 8. The characteristics that scientists use in cladistics are

    a. analogous structures.
    b. shared derived traits.
    c. convergent structures.
    d. shared homologous traits.

9. Bird wings and insect wings are
   a. homologous traits.
   b. derived traits.
   c. analogous traits.
   d. phylogenetic traits.

10. The biological species concept cannot be applied to
   a. species that can produce fertile hybrids.
   b. all bacteria.
   c. species that reproduce asexually.
   d. All of the above

11. Scientific names
   a. must have three Latin words and correct Latin grammar.
   b. include the genus and family.
   c. have rules established by British and American biologists.
   d. enable biologists to communicate regardless of their native language.

12. Which of the following lists the seven classification levels in proper descending order?
   a. kingdom, phylum, class, order, family, genus, species
   b. kingdom, phylum, order, class, family, genus, species
   c. kingdom, phylum, family, class, order, genus, species
   d. phylum, kingdom, class, order, family, genus, species

13. The scientific naming system requires
   a. both words to be underlined or italicized.
   b. the genus to be capitalized.
   c. the species to be the second word.
   d. All of the above

14. Two species that interbreed and produce fertile offspring are
   a. closely related.
   b. incorrectly classified as two different species.
   c. reproductively isolated.
   d. hybrids.

15. Evolutionary systematics is more useful than cladistics when scientists
   a. know how a trait affects the life of an organism.
   b. do not have a lot of information about the organism.
   c. want to be more objective in their analysis.
   d. cannot define an out-group.

• • • • • • • • • • • • • • •

Circle T *if the statement is true or* F *if it is false.*

T F 16. In addition to the seven major classification levels, more than 30 other taxonomic categories are recognized.

T F 17. Convergent evolution leads to shared homologous characters.

T F 18. Classes with similar characteristics are grouped into the same order.

T F 19. Using Linnaeus's system, scientists have been able to name nearly all of the species on Earth.

T F 20. Analogous characters are not inherited from a common ancestor.

T F 21. Establishing evolutionary relationships based on similar traits can be misleading.

T F 22. The biological species concept is the only way to identify different species accurately.

Complete each statement by writing the correct term or phrase in the space provided.

23. The naming system developed by Linnaeus is called _____

_____ .

24. One genus can include several _____ .

25. The six kingdoms of living organisms are Archaebacteria,

_____ , Protista, _____ ,

_____ , and Animalia.

26. Ernst Mayr developed the concept that a(n) _____

_____ is reproductively isolated from other groups.

27. When _____ _____ are incomplete, closely related species can produce hybrids.

28. The biological species concept works best for most members of the kingdom

_____ .

29. _____ _____ leads to similar features in organisms that do not share a recent common ancestor.

30. Scientists use evidence of _____ characters to reconstruct evolutionary history.

31. The evolutionary history of a species is called its _____ .

32. The science of naming and classifying organisms is called

_____ .

Read each question, and write your answer in the space provided.

33. Explain the difference between homologous characters and analogous characters. Give an example of each.

_____

_____

_____

_____

_____

_____

34. How does Linnaeus's system of binomial nomenclature benefit scientists?

_____

_____

35. Which classification system would probably be used first if a scientist discovered five unknown plants? Explain.

_____

_____

_____

36. Explain why Mayr's concept of biological species has limited applications.

_____

_____

_____

_____

Questions 37–39 refer to the figure below. The phylogenetic tree shown below indicates the evolutionary relationships for a hypothetical group of modern organisms, labeled 1–5, and their ancestors, labeled A–E.

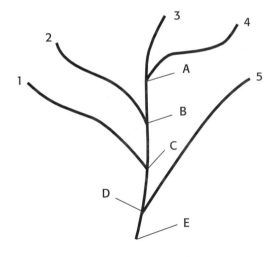

37. Which two modern organisms are likely to be most closely related?

_____

38. What was the most recent common ancestor of organisms *2* and *3*?

_____

39. What was the most recent common ancestor of organisms *1* and *5*?

_____

Name_____ Date _____ Class_____

Classification of Organisms

Complete each statement by writing the correct term or phrase from the list below in the space provided.

analogous character	convergent evolution	kingdom
binomial nomenclature	derived traits	order
biological species	evolutionary systematics	phylogeny
cladistics	family	phylum
cladogram	genus	taxonomy
class		

1. The classification level in which classes with similar characteristics are grouped

   is called a(n) _____ .

2. When taxonomists give varying subjective degrees of importance to characters,

   they are applying _____ _____ .

3. _____ is a system of taxonomy that reconstructs phylogenies by inferring relationships based on similarities derived from a common ancestor.

4. The evolutionary history of a species is its _____ .

5. Orders with common properties are combined into a(n) _____ .

6. Similar families are combined into a(n) _____ .

7. The classification level in which similar genera are grouped is called a(n)

   _____ .

8. A similar feature that evolved through convergent evolution is called a(n)

   _____ _____ .

9. In _____ _____ , organisms evolve similar features independently, often because they live in similar habitats.

10. A(n) _____ is a branching diagram used to show evolutionary relationships.

11. The most general level of classification is _____ .

12. A(n) _____ is a taxonomic category containing similar species.

13. Linnaeus developed a system for naming and classifying organisms, which is called

    _____ .

14. A(n) _____ _____ is a group of interbreeding or potentially interbreeding natural populations that are reproductively isolated from other such groups.

15. Unique characteristics used in cladistics are called _____

_____ .

16. _____ _____ is the two-word system for naming organisms.

Name_____ Date _____ Class _____

Classification of Organisms

Use the list of animals below to complete items 1–3.

bat	frog	horse	rabbit
chicken	goldfish	octopus	spider
eagle	grasshopper	polar bear	whale

1. Organize the animals in the list above into the following four groups of classification. In the space provided below, write the names of the animals from the list above that have each of the described characteristics. Some animals may fall into more than one group.

Group 1

a. Lives on land _____

_____

b. Lives in water _____

_____

Group 2

c. Catches live prey _____

_____

d. Vegetarian _____

_____

Group 3

e. Lays eggs _____

_____

f. Live birth _____

_____

Group 4

g. Lays eggs; has an internal skeleton _____

_____

h. Lays eggs; has an internal skeleton; has feathers _____

_____

i. Live birth; lives on land _____

_____

j. Live birth; lives in water _____

_____

Read each question, and write your answer in the space provided.

2. Which of the four groups above do you think classifies the animals in a manner that best demonstrates their evolutionary relationships? Explain.

_____

_____

_____

_____

_____

_____

_____

3. Are there any animals that did not fit well into any of the categories listed in Groups 1–4? Explain.

_____

_____

_____

_____

Populations

In the space provided, write the letter of the term or phrase that best completes each statement or best answers each question.

_____ 1. The three main patterns of dispersion in a population are
    a. nonrandomly spaced, evenly spaced, and clumped distribution.
    b. nonrandomly spaced, evenly spaced, and unevenly spaced.
    c. randomly spaced, evenly spaced, and clumped distribution.
    d. randomly spaced, evenly spaced, and unevenly spaced.

_____ 2. In the exponential model of population growth, the growth rate
    a. remains constant.   **c.** rises.
    b. declines.   **d.** rises and then declines.

_____ 3. *K*-strategists tend to live in environments that are
    a. unstable.   **c.** unpredictable.
    b. rapidly changing.   **d.** stable and predictable.

_____ 4. The Hardy-Weinberg principle
    a. can predict genotype frequencies.
    b. can predict genetic drift.
    c. applies only to large populations with nonrandom mating.
    d. Both (a) and (b)

_____ 5. In large, randomly mating populations, the frequencies of alleles and genotypes remain constant from generation to generation unless
    a. evolutionary forces are absent.
    b. evolutionary forces act on the population.
    c. the populations are *K*-strategists.
    d. the populations are *r*-strategists.

_____ 6. Natural selection acts on
    a. genotypes.
    b. phenotypes.
    c. both phenotypes and genotypes.
    d. neither phenotypes nor genotypes.

_____ 7. Human height is an example of a
    a. single-gene trait.   **c.** monogenic trait.
    b. double-gene trait.   **d.** polygenic trait.

_____ 8. The range of phenotypes shifts toward one extreme in
    a. stabilizing selection.   **c.** directional selection.
    b. disruptive selection.   **d.** polygenic selection.

_____ 9. Weather and climate are environmental conditions that affect populations and are known as
    a. density-dependent factors.   **c.** logistical factors.
    b. density-independent factors.   **d.** dispersion factors.

T   F   **10.** In a logistical model, exponential growth is limited by a density-dependent factor.

T   F   **11.** Small populations are less likely to be affected by random events, such as natural disasters.

T   F   **12.** The statistical study of populations is called demography.

T   F   **13.** The logistic model of population growth assumes that birth rates and death rates are constant.

T   F   **14.** Populations of species that are *r*-strategists are characterized by exponential growth.

T   F   **15.** Mutation rates in nature are very slow.

T   F   **16.** Inbreeding decreases the proportion of homozygotes in a population.

T   F   **17.** Genetic drift was not a factor in the evolution of humans.

T   F   **18.** In the United States, the allele for sickle cell anemia is slowly increasing in frequency.

T   F   **19.** Natural selection operates efficiently against rare, recessive alleles.

T   F   **20.** Short life spans and many offspring are typical of *r*-strategists.

T   F   **21.** When stabilizing selection eliminates extremes at both ends of a range of phenotypes, the frequencies of the intermediate phenotypes increase.

Complete each statement by writing the correct term or phrase in the space provided.

22. A(n) _____ consists of all of the individuals of a species that live together in one place at one time.

23. Barnacles crowded together on a rock exhibit a type of dispersion called

a(n) _____ _____ .

24. To predict how a population will grow, demographers construct a(n)

_____ of a population, a hypothetical population with the key characteristics of the real population being studied.

25. The population size that an environment can sustain is called the

_____ _____ .

26. _____ tend to produce few offspring that mature slowly.

27. A female robin who chooses a male based on how well he sings is demonstrating

_____ _____ .

28. Migration to or from a population creates _____ flow.

29. If the plot of the phenotypes of a trait in a population is a hill-shaped curve,

the trait exhibits a(n) _____ _____ .

30. A shift in allele frequencies in a population caused by random events or chance

is called _____ _____ .

31. When a recessive allele is present at a frequency of 0.1, only 1 out of 100 individuals will be homozygous recessive and will display the phenotype associated with this allele. However, 18 out of 1,000 individuals will be

_____ and will carry the allele unexpressed.

32. _____ is a type of nonrandom mating that decreases the frequency of heterozygotes in a population.

33. In a(n) _____ growth curve, the growth rate remains the same even though the population size increases steadily.

• • • • • • • • • • • • • •

Read each question, and write your answer in the space provided.

34. If you were determining how the size of a population might change, what features of the population would you examine?

_____

_____

_____

Questions 35–39 refer to the equations below.

$$\textbf{A. } r = \text{birthrate} - \text{death rate}$$

$$\textbf{B. } \Delta N = rN$$

$$\textbf{C. } \Delta N = rN\frac{(K-N)}{K}$$

35. What does r represent in equation A?

_____

36. What number is being calculated in equation B?

_____

37. What type of curve is produced when N in equation B is plotted against time on a graph?

_____

38. What does K represent in equation C?

_____

39. What happens when N approaches K in equation C?

_____

_____

40. Distinguish between *r*-strategists and *K*-strategists.

_____

_____

_____

_____

41. List the five forces that cause populations to evolve.

_____

_____

42. Why does natural selection slowly reduce the frequency of harmful recessive alleles?

_____

_____

_____

_____

Questions 43–44 refer to the figures below.

Graph A

Graph B

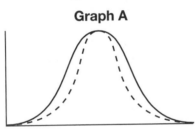 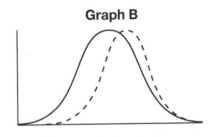

43. What type of distribution does the solid-line curve in each of the graphs above represent?

_____

_____

44. What type of natural selection does the dotted line in Graph A represent? Graph B?

_____

_____

Populations

Use the terms from the list below to fill in the blanks in the following passage.

carrying capacity   genetic drift   population
density-dependent factors   Hardy-Weinberg principle   population density
density-independent factors   *K*-strategists   population model
directional selection   logistic model   population size
dispersion   nonrandom mating   *r*-strategists
exponential growth curve   normal distribution   stabilizing selection
gene flow   polygenic trait

A(n) (1) _____ consists of all the individuals of a species
that live together in one place at one time. Populations tend to grow because
organisms often have multiple offspring over their lifetime. However, limited
resources in an environment eventually limit the growth of a population.

Every population has features that help determine its future. One of the

most important features of any population is its (2) _____

_____ , the number of individuals in a population. Studies
have shown that very small populations are among those most likely to become

extinct. A second important feature is (3) _____

_____ , the number of individuals that live in a given area.
If the individuals in a population are spaced widely apart, they may seldom
encounter one another, making reproduction rare.

A third feature of populations is (4) _____ , which refers
to the way the individuals of the population are arranged in space.

When demographers try to predict how a population will grow, they use a(n)

(5) _____ _____ , a hypothetical population
that exhibits the key characteristics of a real population. A simple population
model describes the rate of population growth as the difference between the
birthrate and the death rate. When the rate of population growth stays the same
and population size is plotted against time on a graph, the population growth

curve resembles a J-shaped curve called a(n) (6) _____

_____ _____ . However, populations do not
usually grow unchecked. The population that an environment can sustain is

called the (7) _____ _____ .

As populations grow, limited resources get used up. These resources are called

(8) _____ _____ _____ because the rate at which they become depleted depends on the density of

the population that uses them. The (9) _____

_____ is a population growth model in which exponential growth is limited by a density-dependent factor.

Many species of plants and insects reproduce rapidly. Their growth is usually limited by environmental conditions, also known as

(10) _____ _____ _____ . Many of these species grow exponentially when environmental conditions

permit their reproduction. Such species are called (11) _____ . Their offspring are small, mature rapidly, and receive little or no parental care. Slow-growing populations, such as whales and redwood trees, are called

(12) _____ because their population density is usually near the carrying capacity (K) of their environment. They are characterized by a long life span, few young, a slow maturing process, reproduction late in life, and extensive parental care.

According to the (13) _____ _____

_____ , the frequencies of alleles in a population do not change unless evolutionary forces act on the population. The evolutionary

forces include the mutation of genes and (14) _____

_____ , which is the movement of alleles into or out of a population. Sometimes individuals prefer to mate with others that live nearby

or are of their own phenotype, a situation called (15) _____

_____ . In small populations, the frequency of an allele can be greatly changed by a chance event, such as a fire or landslide. This change in

allele frequency is called (16) _____ _____ .

Natural selection changes a population through actions on individuals within the population. A trait that is influenced by several genes is called a(n)

(17) _____ _____ . These types of traits tend to exhibit a range of phenotypes clustered around an average value. If you were to plot the height of everyone in your class on a graph, the values would

probably form a hill-shaped curve called a(n) (18) _____

_____ . When selection causes the frequency of a particular

trait to move in one direction, it is called (19) _____

_____ . When selection eliminates extremes at both ends of a range of phenotypes, the frequencies of the intermediate phenotypes increase.

This form of selection is called (20) _____ _____ .

Populations

Use the graph below, which shows the growth of a population over time, to answer questions 1 and 2 below.

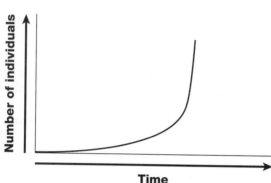

Population Growth

Read each question, and write your answer in the space provided.

1. What type of growth pattern is shown in the graph above? Describe the rate of population growth in this growth pattern.

_____

_____

_____

_____

_____

2. What types of organisms (*r*-strategists or *K*-strategists) are represented by this growth pattern? Describe the features that lead to this growth pattern.

_____

_____

_____

_____

_____

_____

Use the graph below, which shows the frequency of a certain allele in populations of different sizes, to answer questions 3–6 below.

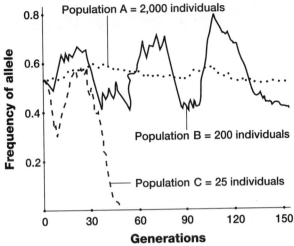

Read each question, and write your answer in the space provided.

3. In which of these populations does the Hardy-Weinberg principle apply? Explain.

_____

_____

_____

_____

4. What happens to the frequency of the allele in the smallest population shown in the graph above? Explain.

_____

_____

_____

_____

5. What conditions must be present in order for the Hardy-Weinberg principle to apply?

_____

_____

_____

6. What factors might cause genotype frequencies to deviate from those predicted by the Hardy-Weinberg equation?

_____

_____

Ecosystems

In the space provided, write the letter of the term or phrase that best completes each statement or best answers each question.

_____ 1. A typical ecosystem might include
    a. large and small mammals.   **c.** birds, trees, and flowers.
    b. microscopic eukaryotes.   **d.** All of the above

_____ 2. Biodiversity is the number of species
    a. of animals living within an ecosystem.
    b. of plants and fungi living within an ecosystem.
    c. of bacteria and protists living within an ecosystem.
    d. living within an ecosystem.

_____ 3. The plants that first grow on an island formed by a volcano are part of a progression called
    a. primary succession.   **c.** secondary succession.
    b. primary productivity.   **d.** the climax community.

_____ 4. In the living portion of the water cycle, water
    a. is retained beneath the surface of Earth as ground water.
    b. evaporates from the soil.
    c. evaporates from dead organisms.
    d. is taken up by the roots of plants.

Questions 5–8 refer to the figure at right.

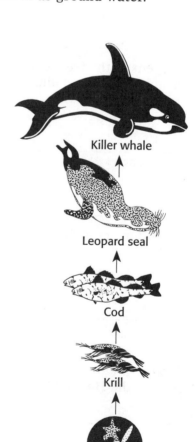

_____ 5. The algae in the figure at right are
    a. decomposers.
    b. consumers.
    c. producers.
    d. herbivores.

_____ 6. The krill in the figure at right are
    a. decomposers.
    b. consumers.
    c. producers.
    d. detritivores.

_____ 7. The figure at right is called a
    a. food chain.
    b. food web.
    c. pyramid of energy.
    d. trophic level.

Killer whale

Leopard seal

Cod

Krill

Algae

_____ 8. The most likely reason that the figure on page 122 has only five levels is that
   a. pollution probably destroyed all of the higher levels.
   b. no other organisms are powerful enough to kill and eat the killer whale.
   c. too much energy is lost at each level to permit more levels.
   d. there is not enough energy initially present at the first level.

_____ 9. The process of succession varies depending on
   a. the plant species involved.
   b. initial environmental conditions and chance.
   c. pioneer species.
   d. competition between species.

_____ 10. The conversion of nitrate to nitrogen gas is called
   a. assimilation.
   b. ammonification.
   c. nitrification.
   d. denitrification.

. . . .
In the space provided, write the letter of the description that
best matches the term or phrase.

_____ 11. ecology

_____ 12. habitat

_____ 13. community

_____ 14. ecosystem

_____ 15. pioneer species

_____ 16. trophic level

_____ 17. herbivores

_____ 18. carnivores

_____ 19. omnivores

_____ 20. detritivores

_____ 21. decomposers

_____ 22. food chain

_____ 23. food web

a. animals at the second trophic level that eat plants

b. the place where a particular population of a species lives

c. bacteria and fungi that cause decay

d. complex, interconnected paths of energy exchanges in an ecosystem

e. the study of the interactions of living organisms with one another and with their physical environment

f. animals that eat both plants and animals

g. a linear path of energy through the trophic levels of an ecosystem

h. the many species that live together in a habitat

i. an organism's place in an ecosystem based on its source of energy

j. organisms that obtain their energy from organic wastes and dead bodies

k. animals at the third trophic level that eat other animals

l. plants that are the first organisms to live in a new habitat

m. a community and all the physical aspects of its habitat

. . . .
Complete each statement by writing the correct term or phrase
in the space provided.

24. In 1866, the German biologist Ernst Haeckel named the study of how

   organisms fit into their environment, calling it _____ .

25. The physical aspects, or _____ _____ , of an
    ecosystem's habitat include soil, water, and weather.

26. When succession occurs in areas where previous growth has occurred, it is

    called _____ _____ .

27. In a(n) _____ _____ , the amount of energy

    stored at each level determines the width of each block.

28. The amount of energy in a trophic level is more accurately determined by

    measuring the _____ (dry weight of tissue) than the

    _____ of organisms.

29. _____ _____ is the process of combining
    nitrogen gas with hydrogen to form ammonia.

30. _____ is the production of ammonia by bacteria during the
    decay of nitrogen-containing urea.

• • • • • • • • • • • • • •
Read each question, and write your answer in the space provided.

31. What components are included in an ecosystem but not in a community?

    _____

    _____

32. Why are both bacteria and fungi important organisms in an ecosystem?

    _____

    _____

33. Describe the process of primary succession that occurred following the retreat
    of the glacier at Glacier Bay, Alaska.

    _____

    _____

    _____

    _____

    _____

    _____

    _____

34. Why are producers an essential component of an ecosystem?

    _____

    _____

35. Why are energy pyramids never inverted?

_____

_____

36. Trace the cycling of water between the atmosphere and Earth.

_____

_____

_____

37. List the four stages of the nitrogen cycle.

_____

_____

Questions 38 and 39 refer to the figure below, which shows the carbon cycle.

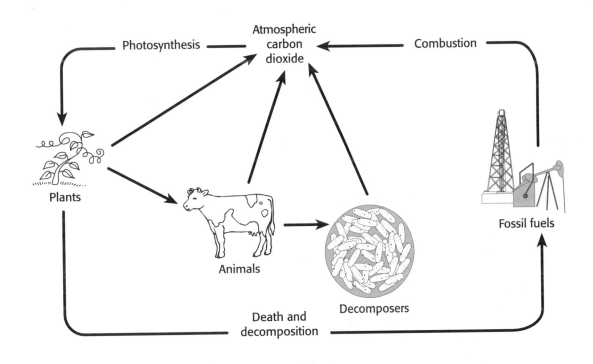

38. How do the living organisms in the figure return carbon atoms to the pool of carbon dioxide in the atmosphere and water?

_____

_____

39. What is the source of the carbon in fossil fuels?

_____

_____

Ecosystems

*Complete each statement by writing the correct term or phrase
from the list below in the space provided.*

abiotic factors	ecology	primary succession
biodiversity	ecosystem	secondary succession
biotic factors	habitat	succession
community	pioneer species	

1. The number of species living within an ecosystem is a measure of its

   _____ .

2. A somewhat regular progression of species replacement is called

   _____ .

3. A(n) _____ consists of a community and all the physical aspects
   of its habitat, such as the soil, water, and weather.

4. The living organisms in a habitat are called _____

   _____ .

5. The first organisms to live in a new habitat are small, fast-growing plants called

   _____ _____ .

6. Succession that occurs where plants have not grown before is called

   _____ _____ .

7. The many different species that live together in a habitat are called a(n)

   _____ .

8. _____ is the study of the interactions of living organisms with one
   another and with their environment.

9. Succession that occurs where previous growth has occurred is called

   _____ _____ .

10. The physical aspects of a habitat are called _____

    _____ .

11. The place where a particular population of a species lives is called its

    _____ .

In the space provided, write the letter of the description that best matches the term or phrase.

_____ 12. primary productivity

_____ 13. producers

_____ 14. consumers

_____ 15. trophic level

_____ 16. food chain

_____ 17. herbivore

_____ 18. carnivore

_____ 19. omnivore

_____ 20. detritivore

_____ 21. decomposers

_____ 22. food web

_____ 23. energy pyramid

_____ 24. biomass

_____ 25. biogeochemical cycle

_____ 26. ground water

_____ 27. transpiration

_____ 28. nitrogen fixation

a. an interconnected group of food chains

b. a pathway formed when a substance enters a living organism, stays for a time in the organism, then returns to the nonliving environment

c. the dry weight of tissue and other organic matter found in a specific ecosystem

d. organisms in an ecosystem that first capture energy

e. water retained beneath the surface of Earth

f. the rate at which organic material is produced by photosynthetic organisms

g. a diagram in which each trophic level is represented by a block with a width proportional to the amount of energy stored in the organisms at that trophic level

h. the process of combining nitrogen with hydrogen to form ammonia

i. organisms that obtain energy by consuming plants or other organisms

j. the evaporation of water from the leaves of plants

k. a level in a graphic organizer based on the organism's source of energy

l. an organism that obtains energy from organic wastes and dead bodies

m. the path of energy through the trophic levels of an ecosystem

n. bacteria and fungi that cause decay

o. an animal that is both an herbivore and a carnivore

p. an animal that eats herbivores

q. an animal that eats plants or other primary producers

CHAPTER
17 **SCIENCE SKILLS: INTERPRETING GRAPHS**

Ecosystems

Use the figure below, which shows the food web of an aquatic ecosystem, to complete items 1–4.

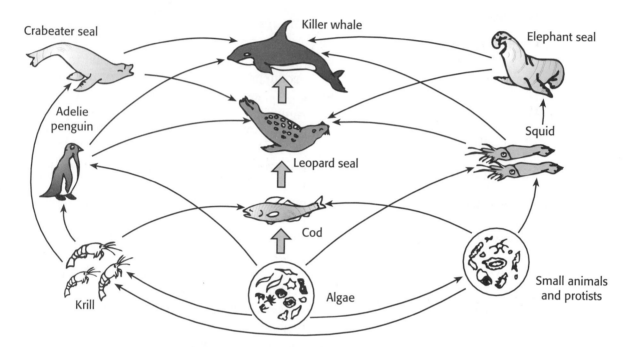

Crabeater seal
Killer whale
Elephant seal
Adelie penguin
Leopard seal
Squid
Cod
Krill
Algae
Small animals and protists

1. In the food web above, there are eight food chains that include krill. In the space provided below, identify all of the organisms in the order in which they occur in four of these eight food chains.

a. Chain 1 _____

_____

b. Chain 2 _____

_____

c. Chain 3 _____

_____

d. Chain 4 _____

_____

Read each question, and write your answer in the space provided.

2. What organisms do cod eat?

_____

_____

3. List all the organisms that eat squid.

_____

_____

4. How many producers are in the food web? Name them.

_____

_____

_____

*Use the figures below, which show trophic levels in an ecosystem,
to complete items 5 and 6 below.*

Done by Counting Organisms at Each Level

First-level carnivores

Herbivores

Marine plankton

A

Done by Weighing Organisms at Each Level

First-level carnivores

Herbivores

Marine plankton

B

Done by Measuring the Calories Stored at Each Level

First-level carnivores

Herbivores

Marine plankton

C

5. Study the three pyramids above. In the space provided, identify which pyramid is the most accurate indicator by writing the correct letter (*A–C*) in the space provided.

_____ a. number of individual organisms

_____ b. measurement of biomass

_____ c. measurement of productivity

6. Which pyramid is the most accurate indicator of the amount of energy available at each trophic level? Explain.

_____

_____

_____

_____

Name _____ Date _____ Class _____

Biological Communities

In the space provided, write the letter of the term or phrase that best completes each statement or best answers each question.

_____ 1. What form of interaction is taking place when a shark devours a seal?
    a. commensalism
    b. mutualism
    c. predation
    d. parasitism

_____ 2. When lions and hyenas fight over a dead zebra, their interaction is called
    a. mutualism.
    b. competition.
    c. commensalism.
    d. parasitism.

_____ 3. Mutualism and commensalism are two types of
    a. symbiosis.
    b. competition.
    c. parasitism.
    d. predation.

_____ 4. All the ways in which a jaguar interacts with its environment make up its
    a. ecosystem.
    b. community.
    c. population.
    d. niche.

_____ 5. In the face of competition, an organism may occupy only part of its fundamental niche. That part is called its
    a. biome.
    b. community.
    c. realized niche.
    d. ecosystem.

_____ 6. The principle of competitive exclusion states that if two species are competing, the species that uses the resource more efficiently will eventually
    a. increase its use of the common resource.
    b. grow exponentially and then level off.
    c. eliminate the other species entirely.
    d. eliminate the other species locally.

_____ 7. Which of the following is NOT part of a freshwater habitat?
    a. profundal zone
    b. tidal zone
    c. littoral zone
    d. limnetic zone

_____ 8. The most important elements of climate are
    a. temperature and weather.
    b. temperature and moisture.
    c. moisture and sun.
    d. rainfall and snowfall.

_____ 9. The greater a community's biodiversity is, the greater its
    a. productivity and stability.
    b. number of niches.
    c. degree of competition.
    d. Both (a) and (b)

_____ 10. A biome's characteristics are affected by
    a. geography.
    b. temperature.
    c. amount of annual rainfall.
    d. All of the above

_____ **11.** Forty percent of all photosynthesis on Earth is accomplished by

    a. bacteria.

    b. coral.

    c. algae.

    d. plankton.

_____ **12.** A major biological community that occurs over a large area of land is called a

    a. biome.

    b. profundal zone.

    c. niche.

    d. population.

• • • • • • • • • • • • •

Circle T *if the statement is true or* F *if it is false.*

T   F   **13.** Back-and-forth evolutionary adjustments between interacting members of an ecosystem are called coevolution.

T   F   **14.** Resources for which species compete include food, nesting sites, living space, light, mineral nutrients, and water.

T   F   **15.** A symbiotic relationship in which both participating parties benefit is called commensalism.

T   F   **16.** The functional role of a particular species in an ecosystem is called its niche.

T   F   **17.** The total niche an organism is potentially able to occupy within an ecosystem is its realized niche.

T   F   **18.** The world's great fisheries are located in shallow ocean waters.

T   F   **19.** When two bacterial species that use different resources were placed in the same culture tube in G. F. Gause's laboratory experiments, one of the species became extinct.

T   F   **20.** Biomes characterized by high annual rainfalls are located at high elevations.

T   F   **21.** Parasites kill their hosts to feed their offspring.

T   F   **22.** Deserts and tundra have similar annual rainfalls.

T   F   **23.** The three major kinds of marine habitats are shallow ocean waters, open sea surface, and deep sea waters.

T   F   **24.** The open, cold plains of the far North make up the biome called taiga.

• • • • • • • • • • • • •

Complete each statement by writing the correct term or phrase in the space provided.

25. _____ often do not kill their prey because they depend on the prey for food, a place to live, and a means to transmit their offspring to new prey.

26. Virtually all plants contain defensive chemicals called _____

_____ .

27. Mild climate and annual precipitation of 75–250 cm favor the growth of the type

of biome called _____ _____

_____ .

28. When two or more species live together in a close, long-term association in which one species benefits and the other is neither harmed nor helped, the

relationship is called _____ .

29. A particular species' _____ is its functional role in an ecosystem.

30. The entire range of conditions an organism can tolerate is its

_____ _____ .

31. Local elimination of one competing species is known as _____

_____ .

32. When sea stars were kept out of experimental plots in the coastal community studied by Robert Paine, the number of species in the ecosystem

_____ .

33. _____ can limit how species use resources.

34. All freshwater habitats are strongly connected to _____ ones.

35. Fewer than 25 cm of precipitation per year falls in two of the world's biomes—

the desert and the _____ .

· · · · · · · · · · · · · ·
Read each question, and write your answer in the space provided.

36. Why is parasitism considered a special case of predation?

_____

_____

_____

_____

37. Explain how the larvae of the cabbage butterfly have overcome the mustard plant's defenses.

_____

_____

_____

_____

38. Explain how predation, competition, and biodiversity are related.

_____

_____

_____

_____

39. Explain how three species of warblers that consume insects in spruce trees can occupy the same forest without violating Gause's principle.

_____

_____

_____

40. Briefly describe the similarities and differences between the limnetic freshwater zone and the surface of the open sea.

_____

_____

_____

_____

41. Where are the world's most abundant fishing grounds located? Explain.

_____

_____

_____

Questions 42–43 refer to the figure below, which shows the results of Gause's experiments with paramecia.

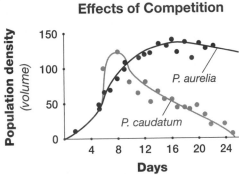

Effects of Competition

42. What principle do these graphs illustrate?

_____

_____

43. What happened to *P. caudatum* when the two species of paramecia were mixed?

_____

_____

CHAPTER

18 VOCABULARY

Biological Communities

In the space provided, write the letter of the description that best matches the term or phrase.

_____ 1. coevolution

_____ 2. predation

_____ 3. parasitism

_____ 4. secondary compounds

_____ 5. symbiosis

_____ 6. mutualism

_____ 7. commensalism

_____ 8. competition

_____ 9. niche

_____ 10. fundamental niche

_____ 11. realized niche

_____ 12. competitive exclusion

_____ 13. biodiversity

a. defensive chemicals used by plants

b. a relationship in which both participating species benefit

c. the entire range of conditions an organism is potentially able to occupy

d. when two species use the same resource

e. back-and-forth evolutionary adjustments between interacting members of an ecosystem

f. two or more species living together in a close, long-term relationship

g. the fundamental role of a species in an ecosystem

h. one organism feeds on and usually lives on or in another larger organism

i. the elimination of a competing species

j. the part of its fundamental niche that a species occupies

k. a relationship in which one species benefits and the other is neither harmed nor helped

l. the variety of living organisms in a community

m. the act of one organism feeding on another

Complete each statement by writing the correct term or phrase in the space provided.

14. The prevailing weather conditions in any given area are called the

_____ .

15. A(n) _____ is a major biological community that occurs over a large area of land.

16. The _____ _____ is a shallow zone near the shore.

17. The _____ _____ is away from the shore but close to the surface.

18. The _____ _____ is a deep-water zone below the limits of effective light penetration.

19. _____ are small organisms that drift in the upper waters of the ocean.

CHAPTER

18 **SCIENCE SKILLS:** INTERPRETING MAPS/INTERPRETING TABLES

Biological Communities

Temperature and moisture tend to vary with latitude and elevation.
These factors play a large role in the biological communities found on
Earth. With this information in mind, complete the following exercises.

*Use the map below, which shows the major terrestrial biomes of
North America and Central America, to complete item 1 below.*

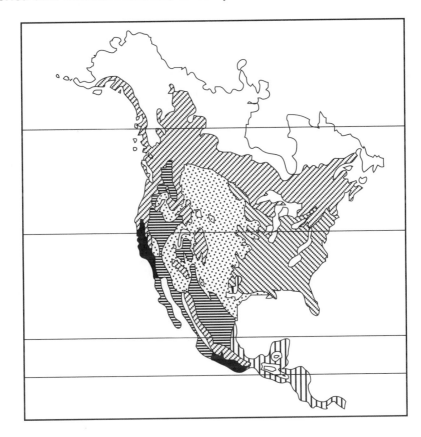

1. In the space provided, write the correct name of each biome next to its key.

a. ☐ _____

b. ▨ _____

c. ▧ _____

d. ⬚ _____

e. ▤ _____

f. ▥ _____

g. ■ _____

Use the table below to answer questions 2 and 3.

Biome	Climate	Annual precipitation	Animal life	Vegetation
Tundra	Brief summer, long winter	< 25 cm	Caribou, ducks	Dwarf willows
Taiga	Brief summer, long winter	35–75 cm	Moose, elk	Firs, spruce
Temperate grasslands	Moderate	25–75 cm	Bison, antelope	Grasses
Temperate deciduous forests	Warm summer, cold winter	75–250 cm	Deer, bears	Birches, maples, shrubs, herbs
Savannas	Seasonal drought, rainy season	90–150 cm	Large herds of grazing animals	Grass with widely spaced trees
Deserts	Moisture varies year to year	< 25 cm	Tortoises, jackrabbits	Sparse vegetation
Tropical rain forests	Rain falls evenly	200–450 cm	More species than any other biome	Tropical plants, trees

Read each question, and write your answer in the space provided.

2. Which two biomes have the least amount of annual precipitation? What is the relationship between the annual precipitation of these biomes and the vegetation they can support?

_____

_____

_____

_____

3. How are the temperature and available moisture of a biome related to the biome's distance from the equator?

_____

_____

_____

_____

Human Impact on the Environment

In the space provided, write the letter of the term or phrase that best completes each statement or best answers each question.

_____ 1. When the sulfur in the atmosphere combines with water vapor, the result is
a. ozone.
b. chlorofluorocarbons.
c. acid rain.
d. ultraviolet radiation.

_____ 2. Chlorofluorocarbons, or CFCs, destroy ozone because ultraviolet radiation breaks the bonds in CFCs and the
a. free chlorine atoms become more stable.
b. free chlorine atoms react to destroy ozone.
c. CFCs prevent ultraviolet radiation from reaching Earth's surface.
d. CFCs are released into the atmosphere.

_____ 3. Global levels of carbon dioxide, methane, and nitrous oxide are
a. rising.
b. remaining constant.
c. falling.
d. too low to be measured accurately.

_____ 4. All of the following are considered nonreplaceable resources EXCEPT
a. topsoil.
b. wood.
c. ground water.
d. animal and plant species.

_____ 5. Plant and animal species are becoming extinct primarily because of
a. disease.
b. hunting.
c. pollution.
d. loss of habitat.

_____ 6. If current birthrates and death rates remain constant, the world's population will double in
a. 20 years.
b. 30 years.
c. 40 years.
d. 60 years.

_____ 7. In which of the following countries is population growth most rapid?
a. United States
b. Nigeria
c. Australia
d. Japan

_____ 8. Worldwide efforts to reduce pollution include all of the following EXCEPT
a. severe restrictions on the use of DDT.
b. taxation and legislation.
c. international agreements to stop CFC production.
d. an international agreement to close all coal-burning facilities.

_____ 9. The first stage of addressing an environmental problem is
a. assessment.
b. risk analysis.
c. public education.
d. political action.

Circle T *if the statement is true or* F *if it is false.*

T   F   **10.** Areas in the northeastern United States have been seriously affected by acid rain because they are downwind from coal-burning plants in the Midwest.

T   F   **11.** In 1985, ozone levels in the atmosphere over Antarctica seemed to be 30 percent higher than the levels found 10 years earlier.

T   F   **12.** Chlorofluorocarbons are stable, harmless heat exchangers.

T   F   **13.** Sources of excess carbon dioxide in the atmosphere include burning of fossil fuels and vegetation.

T   F   **14.** The chemical bonds in carbon dioxide molecules absorb solar energy, trapping heat within the atmosphere.

T   F   **15.** Pollution is no longer considered a major problem largely because of increased public concern and stringent regulation in industrialized and developing countries.

T   F   **16.** The United States has lost one-fourth of its topsoil since 1950.

T   F   **17.** Once pollution enters ground water, there is no effective means of removing it.

T   F   **18.** Chemical pollutants can damage wildlife and pollute ground water.

T   F   **19.** Latin American countries have lower population growth rates than do the United States or Canada.

T   F   **20.** Water withdrawn from aquifers is quickly replaced.

T   F   **21.** Even with information obtained by scientific analysis, it is impossible to predict the consequences of environmental intervention.

Questions 22–26 refer to the figure below.

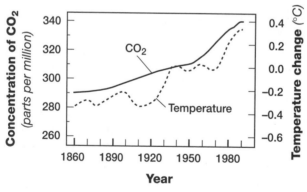

Amount of Carbon Dioxide in the Atmosphere

T   F   **22.** The graph above shows the effects of acid rain.

T   F   **23.** Carbon dioxide concentrations in the atmosphere have more than doubled in the past 100 years.

T   F   **24.** Based on the information in the graph above, the concentration of carbon dioxide in the atmosphere is associated with the global temperature.

T   F   **25.** The warmest years since 1860 occurred after 1980.

T   F   **26.** The temperature fell between 1920 and 1940.

Complete each statement by writing the correct term or phrase in the space provided.

27. When the pH of precipitation in the northeastern United States was measured in 1989, it was almost 100 times as _____ as that of the rest of the United States.

28. In the United States and Canada, forests are being damaged because _____ _____ absorbed by the soil has affected nutrient absorption by the trees.

29. Because of the current condition of the ozone layer, more _____ _____ is reaching Earth's surface.

30. The insulating effect of various gases in Earth's atmosphere is known as the _____ _____ .

31. The increase in global temperatures is called _____ _____ .

32. Examples of chemical pollutants released into the global ecosystem by the agriculture industry are _____ , _____ , and _____ .

33. Since 1650, the human _____ has remained constant, and the _____ _____ has fallen steadily.

34. In the United States, the population _____ _____ is less than half the global rate.

35. Washing cars and watering lawns less often and using efficient faucets are ways to conserve _____ _____ .

36. The Clean Air Act of 1990 requires that power plants install _____ on their smokestacks.

37. The knowledge of _____ is the essential tool that a citizen must have in order to contribute to solving our environmental problems.

Read each question, and write your answer in the space provided.

38. What causes acid rain?

_____

_____

_____

39. How does the presence of the ozone layer affect life on Earth?

_____

_____

_____

40. Explain the relationship between the greenhouse effect and global warming.

_____

_____

_____

41. Explain biological magnification.

_____

_____

_____

_____

42. What has happened to the human death rate in the past several hundred years? Explain.

_____

_____

_____

43. What two approaches have been most effective in reducing pollution in the United States?

_____

_____

_____

44. List the five components necessary to solve an environmental problem successfully.

_____

_____

_____

_____

CHAPTER

19 VOCABULARY

Human Impact on the Environment

Complete each statement by writing the correct term or phrase in the space provided.

1. When sulfur combines with water vapor to form sulfuric acid, the resulting

   precipitation is called _____ _____ .

2. The major cause of ozone destruction is a class of chemicals, invented in

   the 1920s, called _____ .

3. The warming of the atmosphere that results from greenhouse gases is known as the

   _____ _____ .

4. As molecules of chlorinated hydrocarbons pass up through the trophic levels of
   the food chain, they become increasingly concentrated. This process is called

   _____ _____ .

5. _____ are porous rock reservoirs for ground water.

In the space provided, write the letter of the description that best matches the term or phrase.

_____ 6. global change

_____ 7. ozone hole

_____ 8. malignant melanoma

_____ 9. greenhouse gases

_____ 10. chlorinated hydrocarbons

_____ 11. carcinogen

a. a class of compounds that includes
   DDT, chlordane, lindane, and dieldrin

b. examples include acid rain and ozone
   destruction

c. a zone in the atmosphere with a
   below-normal concentration of ozone

d. a potentially lethal form of skin
   cancer

e. a cancer-causing agent

f. gases with heat-trapping ability

Name_____ Date _____ Class _____

Human Impact on the Environment

Use the map below, which shows the distribution of acid rainfall in the United States, to answer questions 1–5.

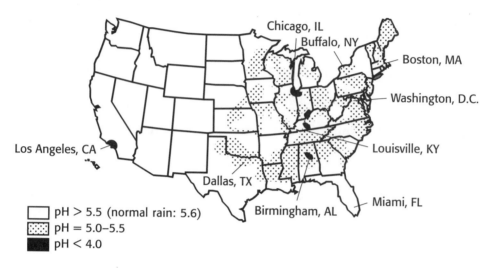

pH > 5.5 (normal rain: 5.6)
pH = 5.0–5.5
pH < 4.0

Read each question, and write your answer in the space provided.

1. How is acid rain formed?

_____

_____

_____

_____

2. What areas shown in the map above experience the most acidic rain?

_____

_____

_____

_____

3. In what region of the United States does most acid rain fall? What conclusions can you draw from this fact?

_____

_____

_____

_____

4. Weather patterns in North America generally move from west to east. Where do you think most of the chemicals that cause acid rain are generated? Explain.

_____

_____

_____

_____

5. Normal rainwater has a pH of approximately 5.6. In lakes and ponds with a pH of 5, salamanders and other amphibians often develop deformities or die. Based on the map on page 142, which section of the United States should have the most abundant and healthiest populations of amphibians?

_____

_____

_____

_____

Name_____ Date _____ Class_____

Introduction to the Kingdoms of Life

*In the space provided, write the letter of the term or phrase that
best completes each statement or best answers each question.*

_____ 1. In terms of cell type, four of the six kingdoms are eukaryotic, while the
other two are

    a. heterotrophic.   **c.** photosynthetic.
    b. multicellular.   **d.** prokaryotic.

_____ 2. One difference between archaebacteria and eubacteria is that the cell walls
of archaebacteria

    a. do not contain peptidoglycan.   **c.** contain more lipids.
    b. contain peptidoglycan.   **d.** do not contain lipids.

_____ 3. A collection of cells that come together for a period of time and then
separate is called

    a. a colony.   **c.** multicellular.
    b. an aggregate.   **d.** protists.

_____ 4. Which of the following statements about protists is NOT true?

    a. Protists are eukaryotes that are neither fungi, plants, nor
animals.
    b. All single-celled prokaryotes are protists.
    c. Some protists use flagella to move.
    d. Many protists reproduce sexually and asexually.

_____ 5. Amoebas are protists that move using extensions of cytoplasm called

    a. flagella.   **c.** pseudopodia.
    b. cilia.   **d.** spores.

_____ 6. Fungi exist in a form made of strings of connected fungal cells called

    a. septa.   **c.** cilia.
    b. spores.   **d.** hyphae.

_____ 7. Most animals and plants have specialized cells that are organized into

    a. organ systems.
    b. tissues.
    c. nerves and muscles.
    d. All of the above

_____ 8. Unlike the cells of organisms in any other kingdom, plant cells have cell
walls composed of

    a. cellulose.   **c.** hyphae.
    b. chitin.   **d.** peptidoglycan.

_____ 9. A type of plant tissue that transports water and dissolved nutrients
is called

    a. vascular tissue.   **c.** nerve tissue.
    b. spongy tissue.   **d.** muscle tissue.

10. Ninety-nine percent of all animals are
   a. vertebrates.
   b. autotrophs.
   c. invertebrates.
   d. unicellular.

11. Plants are complex, multicellular
   a. heterotrophs.
   b. prokaryotes.
   c. protists.
   d. autotrophs.

• • • • • • • • • • • • • • •
Circle T *if the statement is true or* F *if it is false.*

T   F   **12.** All six kingdoms have a body type that includes organs and tissues.

T   F   **13.** Eukaryotes and prokaryotes are found in all but one of the six kingdoms.

T   F   **14.** Eubacteria include prokaryotes that normally live in and on the human body.

T   F   **15.** Methanogens, thermophiles, and halophiles are all types of eubacteria.

T   F   **16.** Multicellular algae fill the evolutionary gap between unicellular protists and complex multicellular organisms.

T   F   **17.** Photosynthetic protists include algae, diatoms, and slime molds.

T   F   **18.** Fungi and bacteria are the primary decomposers in the biosphere.

T   F   **19.** Multicellularity allows cells to specialize.

T   F   **20.** One important role that plants play in most terrestrial food webs is that they release carbon dioxide into the atmosphere.

T   F   **21.** Plants are not found in the extreme polar regions.

T   F   **22.** The success of animals in different habitats is due, in part, to their function as primary producers.

T   F   **23.** Animals are either detritivores, primary consumers, secondary consumers, or parasites.

T   F   **24.** The most diverse groups of animals are the sponges and the cnidarians.

• • • • • • • • • • • • • • •
Complete each statement by writing the correct term or phrase in the space provided.

25. In some kingdoms, organisms use both _____ and

_____ as means to obtain their nutrition, while organisms in other kingdoms use only one means or the other.

26. One distinguishing characteristic of eubacteria is the _____

_____ sequences of the ribosome proteins and RNA polymerases.

27. Eukaryotes and archaebacteria have genes that are interrupted by nontranslated

segments called _____ .

28. While the plasmodial slime mold comes together temporarily with other slime

mold cells to form a(n) _____ , the species *Volvox* forms a

permanently associated _____ _____ .

29. True multicellularity occurs only in _____ .

30. The kind of protist responsible for the disease malaria is called a(n)

_____ .

31. Of the three phyla of fungi, the fungi that make mushrooms are the

_____ .

32. The members of the plant and animal kingdoms are defined according to the

_____ and _____ of their cells.

33. Specialized cells with a common structure and function make up a(n)

_____ .

34. Animal cells differ from plant cells in that animal cells contain no

_____ _____ .

35. In animals, gametes do not divide by _____ first, as they do in plants, but rather fuse directly with one another to form a(n)

_____ .

• • • • • • • • • • • • • • •
Read each question, and write your answer in the space provided.

36. List the six kingdoms, and indicate whether the organisms in each one are prokaryotic or eukaryotic.

_____

_____

37. Explain why colonial organisms and aggregates are not considered multicellular organisms.

_____

_____

_____

_____

38. How do nonvascular plants differ from vascular plants? Give an example of each.

_____

_____

_____

_____

_____

_____

Questions 39–41 refer to the figure at right, which shows a phylogenetic tree of the six kingdoms.

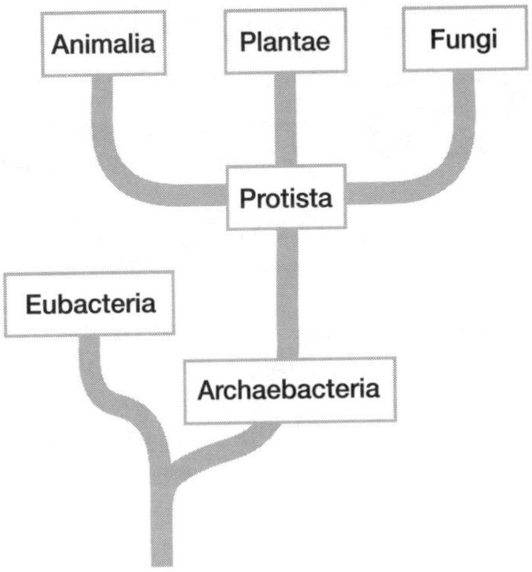

39. In terms of cell type and complexity, explain which two kingdoms are most distantly related.

_____

_____

_____

_____

_____

_____

_____

_____

40. Does the phylogenetic tree reasonably separate prokaryotes from eukaryotes? Explain.

_____

_____

_____

41. Explain why splitting bacteria into two separate kingdoms is justified.

_____

_____

_____

_____

_____

42. List the six categories of animals, and give an example of each.

_____

_____

_____

_____

_____

CHAPTER
20 **VOCABULARY**

Introduction to the Kingdoms of Life

In the space provided, write the letter of the description that best matches the term or phrase.

_____ 1. colonial organism

_____ 2. aggregation

_____ 3. multicellular organism

_____ 4. differentiation

_____ 5. protists

_____ 6. hypha

_____ 7. septa

_____ 8. tissue

_____ 9. organ

_____ 10. organ system

_____ 11. vascular tissue

_____ 12. invertebrates

_____ 13. vertebrates

a. animals that have a backbone

b. a group of specialized cells that transport water and dissolved nutrients

c. a collection of organs that carry out major body functions

d. animals that lack a backbone

e. distinct types of cells with a common structure and function

f. a group of cells that are permanently associated but do not communicate with one another

g. a string of connected fungal cells

h. the process by which cells become specialized in form and function

i. an organism composed of many cells that are permanently associated with one another and that integrate their activities

j. a collection of cells that come together for a period of time and then separate

k. tissues organized into a specialized structure with specific functions

l. the walls dividing one fungal cell from another

m. eukaryotes that are not fungi, plants, or animals

CHAPTER
20 SCIENCE SKILLS: ORGANIZING INFORMATION/INTERPRETING TABLES

Introduction to the Kingdoms of Life

Use the table below to answer questions 1–6.

Kingdom	Cell type	Cell structure	Body type	Nutrition
Eubacteria	Prokaryotic	Peptidoglycan, cell wall	Unicellular	Autotrophic and heterotrophic
Archaebacteria	Prokaryotic	No peptidoglycan, cell wall	Unicellular	Autotrophic and heterotrophic
Protista	Eukaryotic	Mixed	Unicellular and multicellular	Autotrophic and heterotrophic
Fungi	Eukaryotic	Cell wall, chitin	Unicellular and multicellular	Heterotrophic
Plantae	Eukaryotic	Cell wall	Multicellular	Autotrophic
Animalia	Eukaryotic	No cell wall	Multicellular	Heterotrophic

Read each question, and write your answer in the space provided.

1. What characteristics are used to distinguish between the organisms in different kingdoms?

_____

_____

_____

_____

2. What is the primary difference in cell structure between Archaebacteria and Eubacteria?

_____

_____

_____

_____

3. Which kingdoms include multicellular heterotrophic organisms?

_____

_____

_____

4. What primary difference distinguishes the members of kingdoms Archaebacteria and Eubacteria from members of the other kingdoms?

_____

_____

_____

_____

5. What characteristics distinguish fungi from plants?

_____

_____

_____

_____

6. Another possible way to classify organisms would be to separate them into unicellular and multicellular organisms. Explain why this would not be a useful classification system.

_____

_____

_____

_____

_____

CHAPTER 21 TEST PREP PRETEST

Viruses and Bacteria

In the space provided, write the letter of the term or phrase that best completes each statement or best answers each question.

_____ 1. In biologist Wendell Stanley's 1935 investigation of the tobacco mosaic virus, he found that the purified virus
   a. consisted of living organisms that did not retain the ability to infect healthy tobacco plants.
   b. was a crystal that could infect healthy tobacco plants.
   c. was a crystal that could not infect healthy tobacco plants.
   d. consisted of bacteria that could infect healthy tobacco plants.

_____ 2. Each particle of TMV is made of
   a. RNA and proteins.
   b. DNA and proteins.
   c. RNA and lipids.
   d. proteins and lipids.

_____ 3. Viruses are not considered to be living because they do not
   a. maintain homeostasis.
   b. replicate.
   c. metabolize.
   d. All of the above

_____ 4. *Polyhedral virus* refers to the structure of a virus's
   a. nucleic acid.
   b. phage.
   c. capsid.
   d. lipid layer.

_____ 5. HIV can be transmitted
   a. through sexual contact.
   b. through the sharing of nonsterile needles.
   c. to infants during pregnancy or through breast milk.
   d. All of the above

_____ 6. One difference between bacteria and eukaryotes is that
   a. bacterial flagella are more complex than eukaryotic flagella.
   b. bacterial chromosomes are circular, while eukaryotic chromosomes are linear.
   c. bacterial cells are much larger than eukaryotic cells.
   d. many bacteria are multicellular, while all eukaryotes are unicellular.

_____ 7. Bacterial cells lack
   a. chromosomes.
   b. reproductive capability.
   c. flagella.
   d. a cell nucleus.

_____ 8. Which of the following does NOT characterize the structure of *Escherichia coli*?
   a. rigid cell wall
   b. flagella
   c. organelles
   d. pili

_____ 9. In a process called nitrification, chemoautotrophic bacteria that live in the soil play an important role in oxidizing ammonia into
    a. nitrate.
    b. nitrogen gas.
    c. nitrous oxide.
    d. sulfur.

_____ 10. Tuberculosis is a disease of the
    a. brain.
    b. heart.
    c. liver.
    d. lungs.

_____ 11. One way bacteria cause disease is
    a. by metabolizing their hosts.
    b. by producing antibiotics.
    c. through the lysogenic cycle.
    d. None of the above

_____ 12. Which of the following bacterial diseases is NOT transmitted through contaminated water?
    a. cholera
    b. dysentery
    c. bubonic plague
    d. typhoid fever

_____ 13. Mining companies harvest copper or uranium by using
    a. photosynthetic bacteria.
    b. heterotrophic bacteria.
    c. cyanobacteria.
    d. chemoautotrophic bacteria.

• • • • • • • • • • • • • •

In the space provided, write the letter of the description that best matches the term or phrase.

_____ 14. capsid

_____ 15. envelope

_____ 16. glycoproteins

_____ 17. bacteriophage

_____ 18. pathogen

_____ 19. lytic cycle

_____ 20. provirus

_____ 21. lysogenic cycle

_____ 22. bacillus

_____ 23. coccus

_____ 24. spirillum

_____ 25. capsule

_____ 26. antibiotic

_____ 27. conjugation

_____ 28. endospore

_____ 29. heterotrophic bacteria

a. a host chromosome with a viral gene inserted into it

b. proteins with carbohydrate molecules attached

c. a drug that interferes with the life processes in bacteria

d. a rod-shaped bacterial cell

e. bacteria that feed on organic material formed by other organisms

f. a spiral-shaped bacterial cell

g. a thick wall formed around the chromosomes of some bacteria in times of environmental stress

h. a virus's protein coat

i. a cycle in which the viral genome replicates without destroying the host cell

j. a bacteria-infecting virus

k. a cycle of viral infection, replication, and cell destruction

l. a process in which two organisms exchange genetic material

m. a round bacterial cell

n. an agent that causes disease

o. surrounds the capsid of many viruses and helps them enter cells

p. the gel-like layer outside of the cell wall of many bacteria

Complete each statement by writing the correct term or phrase in the space provided.

30. A(n) _____ is a segment of nucleic acids contained in a protein coat.

31. Viruses must rely on _____ _____ for replication.

32. The capsid of viruses may enclose either the nucleic acid _____

or the nucleic acid _____ .

33. Infectious particles called _____ are composed of proteins and have no nucleic acid.

34. HIV gradually infects and destroys so many _____ cells that people with AIDS often die of infections that a healthy immune system would normally resist.

35. The _____ of *E. coli* have two main functions: to adhere to surfaces and to join bacterial cells prior to conjugation.

Questions 36 and 37 refer to the figure below, which shows the human immunodeficiency virus (HIV).

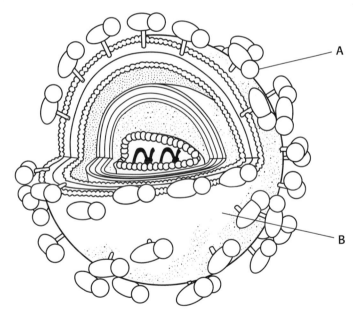

36. The structure labeled *A* is derived from the membrane of the

_____ _____ .

37. The structure labeled *B* is a(n) _____ .

38. In the presence of hydrogen-rich chemicals, _____ bacteria can manufacture all of their own amino acids and proteins.

39. _____ , such as penicillin, work by interfering with different cellular processes of bacteria.

Read each question, and write your answer in the space provided.

40. Describe how HIV reproduces.

41. Name and describe each of the four groups of photosynthetic bacteria.

42. How does *E. coli* reproduce?

43. How was penicillin discovered?

44. List five diseases caused by bacteria.

45. List five viral diseases that are transmitted through person-to-person contact.

— Viruses and Bacteria

Use the terms from the list below to fill in the blanks in the following passage.

bacteriophages   glycoproteins   prions

capsid   lysogenic cycle   provirus

emerging viruses   lytic cycle   viroids

envelope   pathogen   viruses

Segments of nucleic acids contained in a protein coat are called

(1) _____ . The protein coat, or (2) _____

may contain RNA or DNA, but not both. Many viruses have a(n)

(3) _____ , which surrounds the capsid and helps the virus

enter cells. It consists of proteins, lipids, and (4) _____

derived from the host cell. Viruses that infect bacteria are called

(5) _____ .

Any agent that causes disease is called a(n) (6) _____ .

Viruses cause damage when they replicate inside cells many times. When the

viruses break out, the cell is destroyed. The cycle of infection, replication, and

cell destruction is called the (7) _____

_____ .

During an infection, some viruses stay inside the cells but do not make new

viruses. Instead, the viral gene is inserted into the host chromosome and is

called a(n) (8) _____ . Whenever the cell divides, the provirus

also divides, resulting in two infected host cells. This type of replication cycle

is called a(n) (9) _____ _____ .

(continued on next page)

Viruses that evolve in a geographically isolated area and are pathogenic to humans are called (10) _____ _____ .

Infectious disease agents that have a single strand of RNA and have no capsid are called (11) _____ . There is a newly discovered class of infectious particles called (12) _____ , which are composed of protein with no nucleic acid.

In the space provided, write the letter of the description that best matches the term or phrase.

_____ 13. pilus

_____ 14. bacillus

_____ 15. coccus

_____ 16. spirillum

_____ 17. capsule

_____ 18. antibiotics

_____ 19. endospores

_____ 20. conjugation

_____ 21. anaerobic

_____ 22. aerobic

_____ 23. toxins

a. a gel-like layer outside the cell wall and membrane

b. spiral-shaped bacterium

c. bacterial structures that can survive environmental stress

d. an outgrowth on bacteria that attaches to surfaces or other cells

e. round-shaped bacterium

f. a process in which two organisms exchange genetic material

g. oxygen-free environment

h. environment with oxygen

i. chemicals poisonous to eukaryotic cells

j. rod-shaped bacterium

k. chemicals that interfere with life processes in bacteria

CHAPTER
22 TEST PREP PRETEST

Protists

In the space provided, write the letter of the term or phrase that best completes each statement or best answers each question.

_____ 1. Some protists respond to stimuli in their environment using small, light-sensitive organelles called

    a. cilia.   **c.** pseudopodia.
    b. eyespots.   **d.** flagella.

_____ 2. Eukaryotes that lack the features of animals, plants, or fungi are placed in the kingdom

    a. Archaebacteria.   **c.** Protista.
    b. Plantae.   **d.** Animalia.

_____ 3. Protists are found almost everywhere there is

    a. water.   **c.** methane.
    b. carbon monoxide.   **d.** ammonia.

_____ 4. When *Chlamydomonas* reproduces asexually, it divides by mitosis, producing

    a. zygospores.   **c.** haploid gametes.
    b. diploid gametes.   **d.** zoospores.

_____ 5. The green alga *Spirogyra* reproduces sexually by

    a. alternation of generations.   **c.** mitosis.
    b. conjugation.   **d.** aggregation.

_____ 6. Sexual reproduction allows *Chlamydomonas* to delay development of new organisms until

    a. the gametes shed their cell walls.
    b. environmental conditions are favorable.
    c. the zoospores break out of the parent cells.
    d. Both (a) and (b)

_____ 7. Amoebas anchor to nearby surfaces using extensions of cytoplasm called

    a. cilia.   **c.** pseudopodia.
    b. flagella.   **d.** tests.

_____ 8. Because the euglenoid's pellicle is flexible, this organism can

    a. move toward the light.   **c.** reproduce sexually.
    b. survive in a dark environment. **d.** change shape.

_____ 9. *Paramecium* takes in food through its

    a. contractile vacuoles.   **c.** flexible outer pellicle.
    b. cilia-lined oral groove. **d.** two nuclei.

_____ 10. The species of *Plasmodium* that cause the disease malaria are found in

    a. contaminated water.   **c.** the saliva of certain mosquitoes.
    b. infected tsetse flies.   **d.** infected kissing bugs.

_____ 11. Symptoms of giardiasis include
    a. fever and severe heart damage.   **c.** diarrhea, cramps, and vomiting.
    b. fever and lethargy.   **d.** fever, chills, and sweats.

_____ 12. When an individual diatom gets too small because of repeated division, it
    a. grows to full size in its existing shell.
    b. slips out of its shell, grows to full size, and regenerates a new shell.
    c. slips out of its shell, grows to full size, and reinhabits its old shell.
    d. slips out of its shell and lives the rest of its life without a shell.

_____ 13. Algae are distinguished by their cell or body shape and by
    a. the type of photosynthetic pigment they contain.
    b. whether they are unicellular or multicellular.
    c. their method of reproduction.
    d. whether they are heterotrophs or autotrophs.

_____ 14. In addition to causing disease, protists also affect humans through
    a. their role in the nitrogen cycle.
    b. the diseases they transmit to plants.
    c. the diseases they cause in livestock.
    d. All of the above

• • • • • • • • • • • • • • •

In the space provided, write the letter of the description that
best matches the term or phrase.

_____ 15. zygospore

_____ 16. amoebas

_____ 17. forams

_____ 18. green algae

_____ 19. red algae

_____ 20. brown algae

_____ 21. dinoflagellates

_____ 22. zoomastigotes

_____ 23. euglenoids

_____ 24. cellular slime molds

_____ 25. plasmodial slime molds

_____ 26. oomycetes

a. freshwater protists with two flagella; reproduce by mitosis; some are photosynthetic; some are heterotrophic

b. unicellular phototrophs; most have two flagella; most are marine and make up part of plankton; cause red tides

c. individual organisms that behave as separate amoebas; gather together to form slugs during times of environmental stress

d. most are freshwater, unicellular organisms; some are large, multicellular marine organisms

e. a diploid zygote in *Chlamydomonas* with a thick, protective wall

f. water molds, white rusts, and downy mildews that often grow on dead algae and dead animals

g. multicellular organisms found in warm ocean waters; its color results from red photosynthetic pigments

h. marine protists that live in sand or attach to other organisms or rocks

i. a group of organisms that streams along as a mass of cytoplasm

j. multicellular; found mostly in marine environments

k. protists with no cell wall or flagella; live in fresh water, salt water, and soil

l. unicellular heterotrophs that have at least one flagellum; most reproduce asexually

*Complete each statement by writing the correct term or phrase
in the space provided.*

27. Two of the most important features that evolved among the protists are

    _____ _____ and _____ .

28. Some protists have _____ , small organelles containing light-
    sensitive pigments that detect changes in the quality and intensity of light.

29. During conjugation, protists exchange _____

    _____ .

30. *Ulva* is characterized by two distinct multicellular phases: a diploid, spore-

    producing phase called the _____ generation and a haploid,

    gamete-producing phase called the _____ generation.

31. Long, thin projections of _____ extend through the pores in a
    foram's test to aid in swimming and in catching prey.

32. Diatoms can have one of two types of symmetry, _____ or

    _____ .

33. The large brown algae that grow along coasts are known as

    _____ .

34. The stage of *Plasmodium* that lives in mosquitoes and is injected into humans is

    called the _____ ; the second stage of the *Plasmodium* life

    cycle is called the _____ .

*Questions 35–37 refer to the figure at right, which shows a
paramecium.*

35. The structures labeled *A* are _____ ,
    which enable the paramecium to move through the water.

36. The structure labeled *B* is a(n) _____

    _____ .

37. The structure labeled *C* is a(n) _____

    _____ .

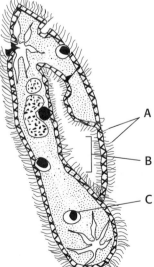

A

B

C

Read each question, and write your answer in the space provided.

38. What diseases caused by protists can be transmitted to humans through drinking water?

_____

_____

39. In what three environments are protists found?

_____

_____

40. Compare the reproductive cycle of *Ulva* with the reproductive cycle of *Spirogyra*. What kinds of protists are *Ulva* and *Spirogyra*?

_____

_____

41. List three of the different types of sexual reproduction in protists.

_____

_____

_____

42. What groups of protists use extensions of cytoplasm for locomotion?

_____

_____

_____

43. What are diatoms, and how are they beneficial?

_____

_____

_____

44. How do diatoms move around?

_____

_____

45. How do people become infected with malaria?

_____

_____

CHAPTER
22 **VOCABULARY**

Protists

Complete the crossword puzzle using the clues provided.

ACROSS

2. _____ of generations: a life cycle characterized by a sporophyte phase and a gametophyte phase.

3. a photosynthetic protist

5. a flexible, cytoplasmic extension

7. a diploid zygote with a thick protective wall

10. the stage of life cycle of *Plasmodium* that infects red blood cells

11. the stage of life cycle of *Plasmodium* that infects the liver

12. a protist that lives in the guts of termites

13. has double shell

14. short flagellum used for movement

DOWN

1. a reproductive cell that produces haploid spores by meiosis

4. protists with no cell walls or flagella

6. a freshwater protist with two flagella

8. a mass of cytoplasm that looks like oozing slime

9. a heterotrophic protist

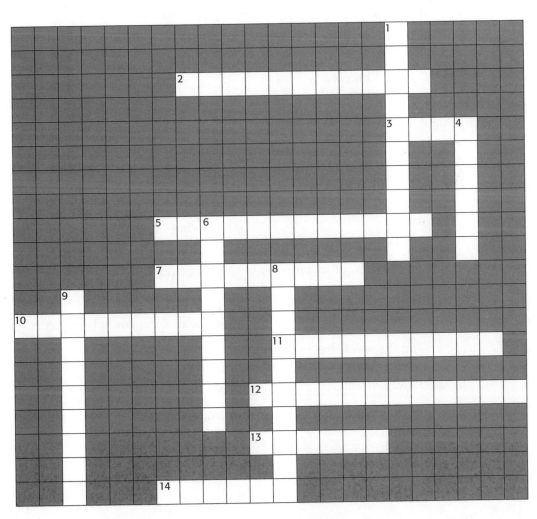

Name ________________________ Date ____________ Class ____________

CHAPTER 23 TEST PREP PRETEST

Fungi

In the space provided, write the letter of the term or phrase that best completes each statement or best answers each question.

______ 1. Which of the following is NOT a characteristic of fungi?
- a. filamentous bodies
- b. cell walls made of chitin
- c. chlorophyll
- d. nuclear mitosis

______ 2. A mycelium helps a fungus absorb nutrients from its environment because it provides
- a. minerals.
- b. a high surface-area-to-volume ratio.
- c. digestive enzymes.
- d. a low surface-area-to-volume ratio.

Questions 3 and 4 refer to the figure at right.

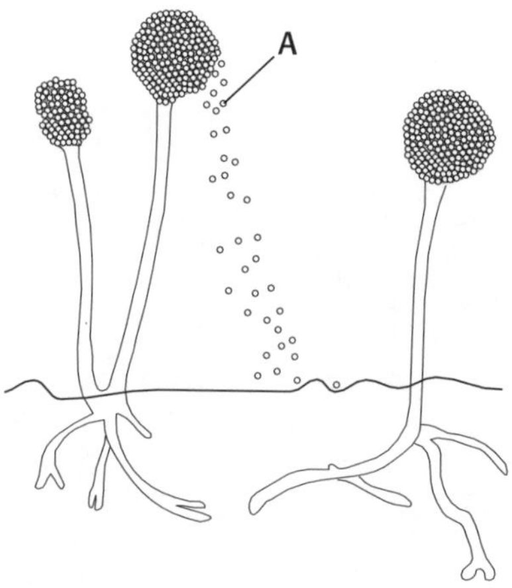

______ 3. The fungus shown at right is a(n)
- a. asomycete.
- b. basidiomycete.
- c. deuteromycete.
- d. zygomycete.

______ 4. The structure labeled *A* in the figure at right is called a
- a. rhizoid.
- b. spore.
- c. stolon.
- d. hypha.

______ 5. Commercial uses for fungi include
- a. antibiotics.
- b. cheese flavorings.
- c. wine-making.
- d. All of the above

______ 6. The classification of organisms in the three phyla of the kingdom Fungi is based on
- a. food.
- b. digestive structure.
- c. cellular structure.
- d. sexual reproductive structure.

______ 7. Most fungal spores are formed by
- a. the fusing of hyphae.
- b. the fusing of asci.
- c. mitosis.
- d. None of the above

______ 8. The recognizable part of *Amanita muscaria* is the
- a. mushroom.
- b. bread mold.
- c. septa.
- d. ascus.

_____ 9. The type of reproduction used by yeasts in which a small cell forms from a large cell and pinches itself off from the large cell is called
   a. fission.
   b. fusion.
   c. budding.
   d. None of the above

_____ 10. In *Amanita muscaria*, the basidia form on the
   a. gills of the mushroom cap.
   b. gills of the stem.
   c. stalk.
   d. mycelium.

• • • • • • • • • • • • • • •

In the space provided, write the letter of the description that best matches the term or phrase.

_____ 11. chitin

_____ 12. hyphae

_____ 13. mycelium

_____ 14. septa

_____ 15. lichen

_____ 16. stolons

_____ 17. mushrooms

_____ 18. zygosporangia

_____ 19. ascus

_____ 20. rhizoid

_____ 21. yeast

_____ 22. basidium

_____ 23. *Aspergillus*

a. a club-shaped sexual reproductive structure after which the phylum Basidiomycota is named

b. members of the phylum Basidiomycota

c. used to ferment soy sauce and produce citric acid

d. tough material found in the cell walls of all fungi

e. a symbiotic association between a fungus and a photosynthetic partner, such as a green alga, a cyanobacterium, or both

f. the saclike reproductive structure for which members of the phylum Ascomycota are named

g. slender filaments that make up the bodies of most fungi

h. common name for unicellular ascomycetes

i. thick-walled sexual structures that characterize members of the phylum Zygomycota

j. tangled mass of hyphae

k. in phylum Zygomycota, the mycelia that grow along the surface of the bread

l. hyphae that anchor the fungus in the bread molds

m. the walls that divide cells in hyphae

• • • • • • • • • • • • • • •

Complete each statement by writing the correct term or phrase in the space provided.

24. Fungi that absorb nutrients in a person's body can cause life-threatening

_____ .

25. The perforated barriers between the cells in some fungi are called

_____ .

26. Fungi exhibit _____ mitosis, in which the nuclear envelope remains intact from prophase to anaphase.

27. When you look at a lichen, you are looking at the _____ , which

   is usually a(n) _____ .

28. Fungi secrete _____ _____ that break down
   organic matter so they can then absorb the decomposed molecules.

29. Sexual reproduction in fungi is initiated when two _____ of
   opposite mating types fuse and form a reproductive structure.

30. The three phyla of fungi form distinctive structures during sexual reproduction.

   Members of the phylum Zygomycota form _____ ; members of

   the phylum Ascomycota form _____ ; members of the phylum

   Basidiomycota form _____ .

31. The unicellular fungi used in the production of baked goods and the production

   of alcoholic beverages are called _____ .

32. Fungal reproductive structures are located at the tips of the

   _____ .

33. Certain fungi play important roles in the nutrition of vascular plants by forming

   symbiotic associations with their roots, called _____ .

34. Biologists use the relative health of _____ and their chemical
   compositions as indicators of air pollutants.

35. Fungi have _____ bodies.

• • • • • • • • • • • • • •
Read each question, and write your answer in the space provided.

36. Give four reasons fungi are no longer classified in the same kingdom as plants.

   _____

   _____

   _____

37. Explain how fungi in the phylum Zygomycota typically reproduce.

   _____

   _____

   _____

   _____

   _____

   _____

38. Distinguish between mitosis in fungi and mitosis in plants and most other eukaryotes.

_____

_____

_____

_____

_____

39. Why can lichen survive in habitats as diverse as polar and desert regions?

_____

_____

_____

40. Give at least two examples of a beneficial and a harmful ascomycete.

_____

_____

_____

_____

41. Describe how _Amanita muscaria_ obtains nutrients.

_____

_____

_____

_____

CHAPTER
23 VOCABULARY

Fungi

In the space provided, write the letter of the description that best matches the term or phrase.

_____ 1. chitin

_____ 2. hyphae

_____ 3. mycelium

_____ 4. zygosporangium

_____ 5. stolon

_____ 6. rhizoid

_____ 7. ascus

_____ 8. yeast

_____ 9. budding

_____ 10. basidium

_____ 11. mycorrhizae

_____ 12. lichen

a. a type of mutualistic relationship formed between fungi and the roots of vascular plants

b. a thick-walled sexual structure

c. the tough polysaccharide found in the hard outer covering of insects and fungal cell walls

d. a symbiosis between a fungus and a photosynthetic partner

e. the hyphae that anchor a fungus to its source of food

f. slender filaments that compose the body of a fungus

g. tangled mass formed by hyphae

h. the mycelia that grow along the surface of a fungus's food

i. a saclike structure in which haploid spores are formed

j. the common name given to unicellular ascomycetes

k. a club-shaped sexual reproductive structure

l. asexual reproduction in which a small cell forms from a larger cell by pinching itself off from the larger cell

CHAPTER
(24) **TEST PREP PRETEST**

Introduction to Plants

In the space provided, write the letter of the term or phrase that best completes each statement or best answers each question.

_____ 1. To reduce water loss, plants have a waxy layer, which covers the nonwoody aboveground parts, called a
    a. mycorrhizae.
    b. cuticle.
    c. stomata.
    d. rhizoid.

_____ 2. In vascular plants, the dominant generation is the
    a. sporophyte.
    b. gametophyte.
    c. bryophyte.
    d. sporangium.

_____ 3. Seeds provide the offspring of plants with all of the following survival advantages EXCEPT
    a. dispersal.
    b. insulation.
    c. nourishment.
    d. delayed growth.

_____ 4. Before flowers, the pollen of the first seed plants was carried by
    a. birds.
    b. animals.
    c. wind.
    d. insects.

_____ 5. The sporophytes of most vascular plants are characterized by all of the following EXCEPT
    a. xylem and phloem.
    b. a shoot.
    c. leaves.
    d. small size.

_____ 6. Nearly all gymnosperms produce gametophytes that develop
    a. within a fruit.
    b. into flowers.
    c. in cones.
    d. in haploid tissue.

_____ 7. The phylum containing the most successful gymnosperms is
    a. Hepatophyta.
    b. Coniferophyta.
    c. Pterophyta.
    d. Ginkgophyta.

_____ 8. Plants in phylum Sphenophyta that have jointed vertical stems and whorls of scalelike leaves are called
    a. ferns.
    b. horsetails.
    c. hornworts.
    d. club mosses.

_____ 9. Which of the following is an important source of starch?
    a. root crops
    b. cassava
    c. potatoes
    d. all of the above

_____ 10. Rayon is made from
    a. wood pulp.
    b. root fibers.
    c. leaves of the foxglove.
    d. tea leaves.

11. Hodgkin's disease is treated with a drug, obtained from the rosy periwinkle, called

    a. salicin.
    b. codeine.
    c. vinblastine.
    d. cortisone.

12. The stems of flax yield a soft, durable fiber that is used to make

    a. cotton.
    b. rayon.
    c. rope.
    d. linen.

Questions 13–16 refer to the figure below, which shows the life cycle of a plant.

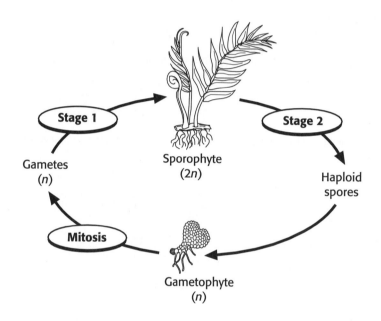

13. What process occurs at stage 1?

    a. mitosis   **c.** fertilization
    b. meiosis   **d.** cell division

14. The structures produced by stage 1 are

    a. spore capsules.
    b. diploid spores.
    c. haploid spores.
    d. zygotes.

15. What process occurs at stage 2?

    a. fertilization   **c.** meiosis
    b. pollination   **d.** mitosis

16. The life cycle above is called

    a. a haploid life cycle.
    b. alternation of generations.
    c. a diploid life cycle.
    d. a phyte life cycle.

In the space provided, write the letter of the description that best matches the term or phrase.

_____ 17. mycorrhizae

_____ 18. stomata

_____ 19. flower

_____ 20. sporophyte

_____ 21. gametophyte

_____ 22. xylem

_____ 23. phloem

_____ 24. seedless vascular plants

_____ 25. Bryophyta

_____ 26. legumes

_____ 27. grain

_____ 28. salicin

a. sporophyte vascular tissue that contains hard-walled, water-conducting cells

b. pores that extend through the cuticle and permit plants to exchange oxygen and carbon dioxide

c. plant phylum that includes true mosses

d. the symbiotic relationships between fungi and the roots of plants

e. a group of plants that have a vascular system, a large sporophyte, and drought-resistant spores

f. a reproductive structure that produces pollen and seeds

g. a pain-relieving chemical found in willow tree bark

h. the dominant stage in the life cycle of nonvascular plants

i. sporophyte vascular tissue that contains soft-walled, sugar-conducting cells

j. the dominant stage in the life cycle of vascular plants

k. plants of the pea family that produce protein-rich seeds in long pods

l. edible dry fruit of a cereal grass

Complete each statement by writing the correct term or phrase in the space provided.

29. Botanists think that _____ may have helped early land plants obtain nutrients from Earth's rocky surface.

30. When specialized cells called _____ _____ change shape, stomata open and close.

31. The part of a plant's body that grows mostly upward is called the

_____ ; the part that grows downward is called the

_____ .

32. In nonvascular plants, the smaller, nongreen _____ depend on

the _____ for nutrients.

33. Seed plants whose seeds do not develop within a fruit are called

_____ .

34. Plants in phylum Pterophyta have coiled young leaves called

_____ and _____ that produce spores on the lower side of fronds.

35. The oldest trees in the world, _____ _____ ,

are _____ , the most successful gymnosperms.

36. Botanically, a(n) _____ is the part of a plant that contains seeds,

and a(n) _____ _____ is any nonreproductive
part of a plant.

37. The most important nonfood products obtained from plants are

_____ and _____ .

38. About 70 percent of the United States corn crop is consumed by

_____ .

39. Paper-making fibers are obtained from wood, cotton, flax, rice,

_____ , and _____ .

• • • • • • • • • • • • • •
Read each question, and write your answer in the space provided.

40. List the three key features of angiosperms, and describe their functions.

_____

_____

_____

41. What is the genetic difference between the gametophyte and the sporophyte?

_____

42. Describe the fundamental differences between vascular and nonvascular plants.

_____

_____

_____

43. List four ways that seeds have influenced the evolution of plants on land.

_____

_____

_____

44. Describe the two types of angiosperms, and list two examples of each.

_____

_____

_____

_____

CHAPTER
(24) VOCABULARY

Introduction to Plants

Write the correct term from the list below in the space next to its definition.

cuticle	nonvascular plant	shoot
embryo	phloem	stoma
flower	root	vascular plant
guard cell	seed	vascular system
meristem	seed plant	xylem

_____ 1. a waxy layer that covers the nonwoody aboveground parts of most plants

_____ 2. permits plants to exchange oxygen and carbon dioxide

_____ 3. one of a pair of specialized cells that open and close the stomata

_____ 4. a system of well-developed vascular tissues

_____ 5. a plant that has no vascular system

_____ 6. a plant with a vascular system

_____ 7. a structure that contains a plant embryo

_____ 8. a vascular plant that produces seeds

_____ 9. a reproductive structure that produces pollen and seeds

_____ 10. tissue made of soft-walled cells that transport organic nutrients

_____ 11. tissue made of hard-walled cells that transport water and mineral nutrients

_____ 12. the part of a plant's body that grows mostly upward

_____ 13. the part of a plant's body that grows mostly downward

_____ 14. zone of actively dividing plant cells

_____ 15. an early stage in the development of a plant or an animal

(continued on next page)

In the space provided, explain how the terms in each pair differ in meaning.

16. rhizoid, rhizome

_____

_____

_____

17. frond, cone

_____

_____

_____

18. gymnosperm, angiosperm

_____

_____

_____

19. fruit, endosperm

_____

_____

_____

20. monocot, dicot

_____

_____

_____

21. vegetative part, vegetable

_____

_____

_____

22. cereal, grain

_____

_____

_____

CHAPTER
24 **SCIENCE SKILLS: COMPARING STRUCTURES/INTERPRETING TABLES**

Introduction to Plants

Use the figures below to complete items 1–3 below.

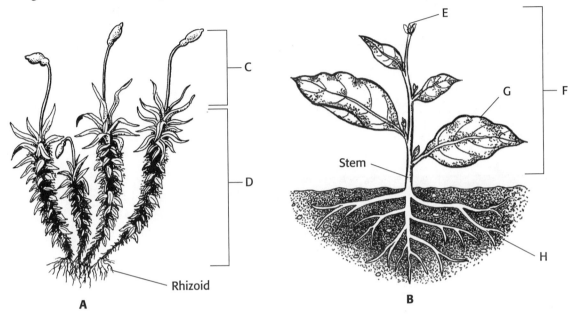

A **B**

1. Study the two plants above. In the space provided, write the letter of the structure (*A–H*) that refers to each term below. Some terms may relate to more than one labeled structure.

_____ **a.** shoot   _____ **f.** meristem

_____ **b.** vascular plant   _____ **g.** undergoes alternation
                                          of generations

_____ **c.** root   _____ **h.** leaf

_____ **d.** sporophyte

_____ **e.** nonvascular plant   _____ **i.** gametophyte

2. Describe two differences between nonvascular plants and vascular plants.

_____

_____

_____

_____

3. What is the basic life cycle of these plants called?

_____

In the tables below, some legumes and grains are rated according to how much of each of four amino acids they contain. A rating of *D* indicates that the food contains a tiny amount of an amino acid. A rating of *A* indicates that the food contains a day's allowance of an amino acid. Use the tables below to answer questions 4–6.

Table 1

Legumes	Amino acids			
	Tryptophan	Isoleucine	Lysine	Cysteine and methionine
Kidney beans	C	B	A	D
Lima beans	C	B	A	C
Soybeans	A	B	A	C
Lentils	C	B	A	D
Navy beans	C	B	A	D

Table 2

Grains	Amino acids			
	Tryptophan	Isoleucine	Lysine	Cysteine and methionine
Wheat	B	C	C	B
Rye	C	C	C	B
Barley	A	C	C	B
Millet	A	C	C	B
Oatmeal	B	C	C	B

Read each question, and write your answer in the space provided.

4. According to Table 1, which legume is the best source of all the amino acids rated?

   _____

   _____

5. According to Table 2, which grain is the best source of all the amino acids rated?

   _____

   _____

6. Compare the ratings of the legumes with those of the grains. Are legumes and grains high in the same amino acids? Explain.

   _____

   _____

   _____

CHAPTER

(25) TEST PREP PRETEST

Plant Reproduction

In the space provided, write the letter of the term or phrase that best completes each statement or best answers each question.

_____ 1. The gametophyte of a nonvascular plant produces sperm in a structure called a(n)
   a. sporangium.
   b. archegonium.
   c. antheridium.
   d. sorus.

_____ 2. Moss gametophytes grow in tightly packed clumps so that when water covers them,
   a. sperm can swim to nearby archegonia and fertilize the eggs.
   b. eggs can float to nearby antheridia.
   c. the spores can germinate.
   d. All of the above

_____ 3. In seedless vascular plants, the archegonia and antheridia develop on the
   a. roots of the gametophytes.
   b. upper surfaces of the gametophytes.
   c. lower surfaces of the sporophytes.
   d. lower surfaces of the gametophytes.

_____ 4. In mosses and ferns, the diploid sporophytes produce spores by
   a. meiosis.
   b. mitosis.
   c. budding.
   d. All of the above

_____ 5. Compared to their gametophytes, the sporophytes of seedless vascular plants are
   a. much smaller.
   b. the same size.
   c. much larger.
   d. variable in size throughout the life cycle.

_____ 6. Reproduction in seed plants differs from that in seedless plants in that
   a. a seed-plant gametophyte is larger than the sporophyte.
   b. a seed-plant gametophyte is part of the sporophyte.
   c. a seed-plant sporophyte does not release spores.
   d. spores are released by the gametophytes of seed plants.

_____ 7. In seed plants, the female gametophyte is found within the
   a. seed.
   b. pollen tube.
   c. male gametophyte.
   d. ovule.

_____ 8. The purpose of cotyledons is to transfer nutrients to the
   a. embryo.
   b. leaves.
   c. gametophytes.
   d. endosperm.

_____ 9. In gymnosperms, the female cones produce
    a. ovules.
    b. pollen.
    c. ovules and seeds.
    d. pollen and seeds.

_____ 10. Some plants reproduce vegetatively using
    a. stolons.
    b. bulbs.
    c. tubers.
    d. All of the above

_____ 11. The process by which two sperm fuse with cells of the female gametophyte to produce both a zygote and endosperm is called
    a. alternation of generations.
    b. meiosis.
    c. double fertilization.
    d. asexual reproduction.

_____ 12. The male reproductive parts of a flower are called
    a. petals.
    b. stamens.
    c. sepals.
    d. pistils.

_____ 13. Kalanchoës reproduce sexually by
    a. stem cuttings.
    b. plantlets.
    c. tiny seeds produced in flowers.
    d. air roots.

_____ 14. Growing African violets from the leaf cuttings of a parent plant is one example of
    a. plant propagation.
    b. grafting.
    c. sexual reproduction.
    d. tissue culture.

• • • • • • • • • • • • • •

Circle T _if the statement is true or_ F _if it is false._

T   F   **15.** In the life cycle of a moss, the gametophyte grows from the sporophyte and remains attached to it.

T   F   **16.** The gametophytes of both mosses and ferns are haploid.

T   F   **17.** The spores of seed plants remain within the tissue of a sporophyte and develop into gametophytes.

T   F   **18.** A pollen tube enables an egg to pass directly to a sperm.

T   F   **19.** Angiosperm pollen grains contain two sperm cells.

T   F   **20.** In conifers, the series of events from pollination to the formation of a seed can take up to 2 years.

T   F   **21.** Wind-pollinated flowers usually open only at night and are pollinated by nighttime visitors, such as bats.

T   F   **22.** Potatoes and caladiums are examples of plants that produce tubers.

In the space provided, write the letter of the description that best matches the term or phrase.

_____ 23. archegonium

_____ 24. antheridium

_____ 25. sorus

_____ 26. pollination

_____ 27. pollen grain

_____ 28. seed coat

_____ 29. sepals

_____ 30. anther

_____ 31. pistil

_____ 32. bulb

a. the innermost whorl of a flower; produces ovules

b. hardened outer cell layers of an ovule

c. the structure of nonvascular plants that produces eggs

d. a cluster of sporangia

e. a type of modified stem with fleshy leaves

f. a pollen-producing sac at the top of a stamen

g. the outermost whorl of a flower

h. the structure of nonvascular plants that produces sperm

i. the male gametophyte of a seed plant

j. the transfer of pollen grains from a male reproductive structure of a plant to a female reproductive structure of a plant

Complete each statement by writing the correct term or phrase in the space provided.

33. Spores are produced in a(n) _____ in mosses. A cluster of

these forms a(n) _____ in ferns.

34. In seed plants, male gametophytes develop into _____

_____ , and female gametophytes develop within the

_____ .

35. In both mosses and ferns, gametophytes produce gametes inside

_____ and _____ .

36. In angiosperms, gametophytes grow from _____ produced by

meiosis within the _____ and _____ of their
flowers.

37. Ferns, irises, and sugarcane have modified stems called _____ ,

which are used in _____ reproduction.

38. A sperm cell fusing with two other haploid cells to form a(n)

_____ cell that develops into _____ is a

characteristic of a(n) _____ life cycle.

39. How are the life cycles of a moss and a fern similar? How are they different?

_____

_____

_____

40. Describe a unique method of vegetative reproduction in kalanchoë.

_____

_____

41. Briefly describe three methods of vegetative plant propagation.

_____

_____

_____

_____

Questions 42–44 refer to the figure below, which shows a fern life cycle.

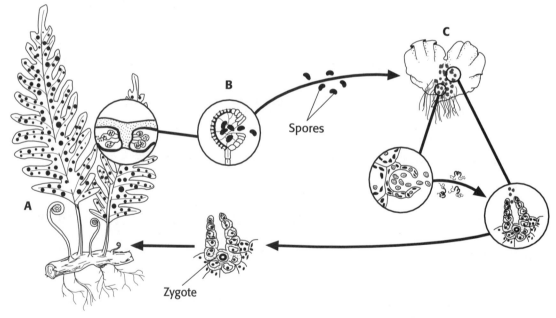

42. What generation is labeled *A*?

_____

43. What is the name of the plant structure labeled *B*?

_____

44. What generation of the fern is labeled *C*?

_____

CHAPTER
25 **VOCABULARY**

Plant Reproduction

Write the correct term from the list below in the space next to its definition.

archegonium   double fertilization   pistil   seed coat
anther   ovary   pollination   sepal
antheridium   ovule   pollen grain   sorus
cotyledon   petal   pollen tube   stamen

_____ **1.** a structure that produces eggs

_____ **2.** a structure that produces sperm

_____ **3.** a cluster of sporangia on a fern frond

_____ **4.** contains a male gametophyte of a seed plant

_____ **5.** the part of the sporophyte in which the female gametophyte develops

_____ **6.** the transfer of pollen grains from the male to the female reproductive structure

_____ **7.** grows from a pollen grain to an ovule

_____ **8.** protects the embryo of a seed from mechanical injury

_____ **9.** leaflike structure that is part of a plant embryo

_____ **10.** one of the flower parts that protects a flower from damage when it is a bud

_____ **11.** one of the flower parts that attract pollinators

_____ **12.** a flower structure that consists of a threadlike filament topped with an anther

_____ **13.** a pollen-producing sac

_____ **14.** a flower structure that produces ovules

_____ **15.** a pistil's swollen lower portion

_____ **16.** two sperm fusing with cells of the female gametophyte to produce both a zygote and an endosperm

(continued on next page)

Complete each statement by writing the correct term or phrase in the space provided.

17. The type of reproduction by which plants produce offspring from vegetative

parts is called _____ _____ .

18. Growing new plants from seeds or from vegetative parts is called

_____ _____ .

19. _____ _____ is a technique used to grow plants
from pieces of plant tissue.

CHAPTER

26   **TEST PREP PRETEST**

Plant Structure and Function

In the space provided, write the letter of the term or phrase that best completes each statement or best answers each question.

_____ 1. The dermal tissue of a vascular plant
  a. conducts water, mineral nutrients, and carbohydrates.
  b. performs photosynthesis.
  c. forms the protective outer layer of the plant.
  d. stores water and carbohydrates.

_____ 2. The primary photosynthetic organs of plants are the
  a. leaves.   c. roots.
  b. stems.   d. flowers.

_____ 3. Which of the following does NOT describe a nonwoody stem?
  a. Pores in the epidermis allow gas exchange.
  b. The stem is usually green and flexible.
  c. The center of the stem is called heartwood.
  d. Vascular tissues are arranged in bundles.

_____ 4. The vascular tissue of a leaf is found in the
  a. petiole.   c. mesophyll.
  b. veins.   d. palisade layer.

_____ 5. When stomata are open, water vapor diffuses out of a leaf in a process called
  a. photosynthesis.   c. osmosis.
  b. germination.   d. transpiration.

_____ 6. Water will keep moving upward in a plant as long as there is an unbroken column of water in the
  a. phloem.   c. roots.
  b. xylem.   d. stomata.

_____ 7. When guard cells take in water, they swell and bend away from each other, opening the
  a. sink.   c. roots.
  b. chloroplasts.   d. stomata.

_____ 8. Translocation is the movement of organic compounds from a source to a
  a. sink.   c. root.
  b. stoma.   d. guard cell.

_____ 9. Which of the following does NOT describe sugar maple leaves?
  a. They have a thin, flattened blade.
  b. Most have five sharply toothed lobes.
  c. They are compound.
  d. They change from light green to yellow, orange, or red in the fall.

_____ **10.** Sugar passes through the xylem of a sugar maple tree as part of a watery solution called

a. maple syrup.
b. sap.

c. heartwood.
d. cork.

• • • • • • • • • • • • • •

In the space provided, write the letter of the description that best matches the term or phrase.

_____ **11.** dermal tissue

_____ **12.** tracheids

_____ **13.** vessel cell

_____ **14.** petiole

_____ **15.** leaflets

_____ **16.** mesophyll

_____ **17.** palisade layer

_____ **18.** node

_____ **19.** internodes

_____ **20.** cortex

_____ **21.** pith

_____ **22.** vascular bundle

_____ **23.** heartwood

_____ **24.** sapwood

_____ **25.** cork

_____ **26.** root hairs

_____ **27.** root cap

a. sections of a compound leaf

b. slender projections from the epidermal cells just behind a root tip that increase the absorption of water and minerals

c. the inner layers of ground tissue in a stem

d. in the center of a woody stem, the part that contains xylem cells that no longer conduct water

e. the ground tissue in a leaf

f. makes up the protective outer layer of a plant

g. in a woody stem, the layer containing xylem cells that still conduct water

h. spaces on a stem between nodes

i. a type of xylem cell with large perforations in its ends

j. a mass of cells that covers an actively growing root tip

k. the ground tissue surrounding the vascular tissue in a root

l. a stalk that attaches a leaf to a stem

m. cluster of vascular tissues in a herbaceous stem

n. the dermal tissue that is located on woody stems and roots and consists of several layers of dead cells

o. the point at which a leaf is attached to a stem

p. narrow, elongated xylem cells with pits through which water flows

q. rows of closely packed, columnar mesophyll cells just beneath the upper epidermis

• • • • • • • • • • • • • •

Complete each statement by writing the correct term or phrase in the space provided.

28. The loss of water by _____ creates a pull that draws water up

through the _____ in the stem and into the leaves.

29. Roots take in water from the soil by _____ .

30. The cells that carry out metabolic functions for the sieve-tube cells of phloem

are called _____ _____ .

31. The _____ inside a sugar maple can be collected and refined for use in the production of maple syrup.

32. Dermal tissue prevents water loss, and it functions in _____

exchange and the absorption of _____ _____ .

33. Herbaceous plants have _____ stems that are

_____ and usually _____ .

34. Ground tissue stores water, _____ , and

_____ , and it supports a plant's _____

_____ .

35. Münch's model of translocation is often called the _____

_____ model.

• • • • • • • • • • • • • •
Read each question, and write your answer in the space provided.

36. Differentiate between nonwoody stems and woody stems.

_____

_____

_____

_____

_____

_____

_____

37. How are the seeds of a sugar maple tree dispersed?

_____

_____

38. Trace the movement of water through a plant.

_____

_____

_____

_____

39. Explain why the opening and closing of stomata by guard cells is an example of homeostasis in action.

_____

_____

_____

40. Identify three reasons that the movement of organic compounds in a plant is more complex than the movement of water.

_____

_____

_____

Questions 41 and 42 refer to the figure below, which shows the pressure-flow model.

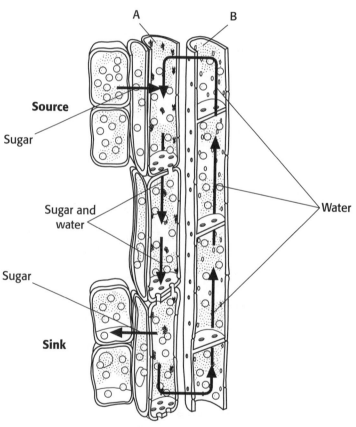

41. Identify the structure labeled *A*. How does sugar enter this structure?

_____

_____

42. Identify the structure labeled *B*. How does water leave this structure?

_____

_____

26 VOCABULARY

Plant Structure and Function

In the space provided, write the letter of the description that best matches the term or phrase.

_____ **1.** dermal tissue

_____ **2.** ground tissue

_____ **3.** epidermis

_____ **4.** cork

_____ **5.** vessels

_____ **6.** sieve tube

_____ **7.** cortex

_____ **8.** root hair

_____ **9.** root cap

_____ **10.** herbaceous plant

_____ **11.** vascular bundle

_____ **12.** pith

_____ **13.** heartwood

_____ **14.** sapwood

_____ **15.** petiole

_____ **16.** mesophyll

a. a plant with stems that are flexible and usually green

b. ground tissue surrounding the vascular bundles in a stem

c. conducting strands in xylem

d. the wood in the center of a mature stem or tree trunk

e. a stalk that attaches a leaf to a stem

f. one of the slender projections of epidermal cells just behind a root tip

g. the ground tissue in a leaf

h. a bundle of xylem and phloem in vascular plants

i. lies outside the heartwood and contains vessel cells that can conduct water

j. dermal tissue in the nonwoody parts of a plant

k. a conducting strand in phloem

l. makes up much of the inside of the nonwoody parts of a plant

m. dermal tissue on woody stems and roots

n. forms the protective outer layer of a plant

o. a cell mass that covers and protects an actively growing root tip

p. ground tissue in the center of a stem or root

(continued on next page)

Complete each statement by writing the correct term or phrase in the space provided.

17. The loss of water vapor from a plant is called _____ .

18. The term _____ refers to a part of a plant that provides organic compounds for other parts of the plant.

19. The term _____ refers to a part of a plant that organic compounds are delivered to.

20. _____ is the movement of organic compounds within a plant from a source to a sink.

CHAPTER
27 **TEST PREP PRETEST**

Plant Growth and Development

In the space provided, write the letter of the term or phrase that best completes each statement or best answers each question.

_____ 1. Apical meristems located at the tips of stems and roots produce
    a. primary growth.   **c.** tertiary growth.
    b. secondary growth.   **d.** no growth.

_____ 2. Secondary growth is produced by cell division in the cork cambium and the
    a. apical meristem.   **c.** apical cambium.
    b. bark.   **d.** vascular cambium.

_____ 3. To complete its life cycle, a biennial plant takes
    a. one growing season.
    b. two growing seasons.
    c. three growing seasons.
    d. more than three growing seasons.

_____ 4. The flowers of bread wheat, like all grass flowers, lack
    a. pistils and stamens.   **c.** stamens and sepals.
    b. pistils and sepals.   **d.** petals and sepals.

Questions 5–7 refer to the figures below.

Corn seed

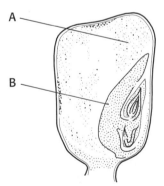

Bean seed

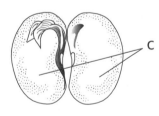

_____ 5. The structure labeled *A* is the
    a. embryo.   **c.** endosperm.
    b. cotyledon.   **d.** root.

_____ 6. The structure labeled *B* is the
    a. embryo.   **c.** endosperm.
    b. cotyledon.   **d.** root.

_____ 7. The structures labeled *C* are the
    a. cotyledons.   **c.** endosperm.
    b. embryos.   **d.** roots.

_____ 8. In a kernel of bread wheat, the ovary wall and seed coat are part of the

a. bran.
b. wheat germ.
c. embryo.
d. aleurone.

_____ 9. Which of the following is NOT a reason that plants need potassium?

a. photosynthesis
b. active transport
c. osmotic balance
d. enzyme activation

_____ 10. The major difference between plant and animal development is that

a. animals continue to develop even after they become adults.
b. plants continue to develop throughout their lives.
c. only plants have genes that guide their development.
d. some animal cells can reverse their development.

_____ 11. All of the following are major mineral nutrients needed by plants EXCEPT

a. phosphorus.
b. sulfur.
c. calcium.
d. zinc.

_____ 12. The Dutch biologist Frits Went showed that the bending of plants toward light is caused by a chemical called

a. auxin.
b. agar.
c. culm.
d. ethylene.

_____ 13. A tropism is a growth response

a. toward light.
b. to touch.
c. toward or away from a stimulus.
d. toward gravity.

_____ 14. Many of a plant's responses to environmental stimuli are caused by

a. the length of the nights.
b. hormones.
c. temperature.
d. All of the above

• • • • • • • • • • • • • •

Circle T *if the statement is true or* F *if it is false.*

T F 15. The process of germination occurs when a plant embryo resumes its growth.

T F 16. Some seeds cannot sprout until their seed coats are damaged.

T F 17. In a tissue culture, cells divide and then undergo differentiation.

T F 18. Trees, shrubs, and woody vines that drop all of their leaves at the end of the growing season each year are classified as annuals.

T F 19. Primary growth results in an increase in a plant's width.

T F 20. If there is not enough air in the spaces between soil particles, there may not be enough oxygen available for a plant's roots and the plant could die.

T F 21. The nutrient nitrogen is a part of all proteins, nucleic acids, chlorophylls, and coenzymes in plants.

T F 22. The inhibition of bud growth along the stem by auxin is called apical dominance.

T F 23. A thigmotropism is a response to gravity.

T F 24. In many dicots, the shoot of a seedling is covered by a protective sheath.

Complete each statement by writing the correct term or phrase in the space provided.

25. Once water has entered the seed and the seed coat breaks, the seedling will

begin to grow if adequate _____ and _____ are available.

26. The plant tissues that result from primary growth are known as

_____ _____ .

27. Wheat grains are high in _____ , a sticky mixture of proteins that makes bread dough elastic.

28. _____ _____ are usually formed every year by thick layers of secondary xylem.

29. Virtually all annuals are _____ plants, and they complete their

life cycle within _____ growing season(s).

30. Apical meristems are located at the _____ of

_____ and _____ .

31. _____ is a gaseous compound that _____ fruit ripening and loosens the fruit of cherries, blackberries, and blueberries.

32. A shoot that grows up out of the ground shows both positive

_____ and negative _____ .

33. _____ is a condition in which a seed or a plant remains inactive even when conditions are suitable for growth.

Read each question, and write your answer in the space provided.

34. What must happen before a seed can germinate?

_____

_____

_____

35. Explain how woody stems increase in width. What is this growth called?

_____

_____

_____

_____

_____

36. Explain how the apical meristems of wheat plants protect them from grazing animals.

_____

_____

_____

_____

37. How can new plants be grown in a tissue culture?

_____

_____

_____

38. What are the three raw materials needed by plants in the greatest amounts, and how are they used?

_____

_____

39. List at least three mineral nutrients that are necessary for plant growth.

_____

40. Summarize how Frits Went demonstrated the presence of the chemical auxin in a shoot tip.

_____

_____

_____

_____

41. How can plants be categorized based on their response to the length of days and nights?

_____

_____

_____

_____

42. Explain how dormancy benefits a seed or plant.

_____

_____

_____

_____

Name_____ Date _____ Class _____

Plant Growth and Development

Write the correct term from the list below in the space next to its definition.

annual   cork cambium   secondary growth
annual ring   germination   vascular cambium
apical meristem   perennial
biennial   primary growth

_____ 1. a plant that lives for several years

_____ 2. growth that increases the length or height of a plant

_____ 3. the meristem that lies within the bark

_____ 4. region where primary growth is produced

_____ 5. plant that takes two growing seasons to complete its life cycle

_____ 6. the meristem that lies just under the bark

_____ 7. layer of secondary xylem formed each year

_____ 8. a plant that completes its life cycle and dies within one growing season

_____ 9. growth that increases the width of a plant's stems and roots

_____ 10. when a plant embryo resumes its growth

In the space provided, write the letter of the description that best matches the term or phrase.

_____ 11. mineral nutrient

_____ 12. auxin

_____ 13. hormone

_____ 14. apical dominance

_____ 15. tropism

_____ 16. photoperiodism

_____ 17. dormancy

a. a response in which a plant grows either toward or away from a stimulus

b. condition in an inactive seed or plant

c. the inhibition of the growth of buds along a stem by the apical meristem

d. the response of plants to the length of days and nights

e. a chemical that causes a stem to bend toward light

f. needed by plants in small amounts

g. a chemical produced in one part of an organism and transported to another part, where it causes a response

Introduction to Animals

In the space provided, write the letter of the term or phrase that best completes each statement or best answers each question.

_____ 1. The cells of all animals are organized into structural and functional units called tissues EXCEPT for the cells of

   a. sponges.   **c.** flatworms.
   b. cnidarians.   **d.** roundworms.

_____ 2. Animals that have their body parts arranged around a central point are

   a. asymmetrical.   **c.** bilaterally symmetrical.
   b. radially symmetrical.   **d.** spherically symmetrical.

_____ 3. Sponges digest their food

   a. in the coelom.   **c.** extracellularly.
   b. in a gastrovascular cavity.   **d.** intracellularly.

_____ 4. All animals EXCEPT sponges and single-celled organisms digest their food

   a. in the coelom.   **c.** extracellularly.
   b. in a one-way gut.   **d.** intracellularly.

_____ 5. An animal in which the space between the body wall and gut is completely filled with tissues and organs is called a(n)

   a. acoelomate.   **c.** coelomate.
   b. pseudocoelomate.   **d.** vertebrate.

_____ 6. An animal whose gut has only one opening has a(n)

   a. intervascular cavity.   **c.** specialized digestive tract.
   b. gastrovascular cavity.   **d.** one-way digestive system.

_____ 7. A true coelom

   a. occurs in radially symmetrical animals.
   b. is located within the endoderm.
   c. protects internal organs from the movement
      of surrounding muscles.
   d. is composed of mesoderm.

_____ 8. In an open circulatory system, the route of the blood is

   a. heart, blood vessels, tissues, heart.
   b. heart, open spaces, body cavity, tissues, heart.
   c. heart, blood vessels, heart.
   d. heart, blood vessels, body cavity, tissues, open spaces, heart.

_____ 9. Parthenogenesis is a method of

   a. sexual reproduction.   **c.** asexual reproduction.
   b. fragmentation.   **d.** fertilization.

_____ 10. Specialized areas for food storage and chemical digestion are found in a(n)

   a. excretory system.   **c.** gastrovascular cavity.
   b. one-way digestive system.   **d.** coelom.

_____ 11. The most accurate method of determining the evolutionary relationships of different animal species is by
  a. comparing their fossils.   c. comparing their size.
  b. looking at a phylogenetic tree.   d. comparing their DNA.

• • • • • • • • • • • • • •
Circle T *if the statement is true or* F *if it is false.*

T F **12.** Asexual reproduction requires the union of male and female gametes.

T F **13.** The bodies of sponges completely lack symmetry.

T F **14.** All animals have tissues.

T F **15.** Most bilaterally symmetrical animals have evolved a definite head end through a process called cephalization.

T F **16.** A pseudocoelom is a body cavity located between the endoderm and the mesoderm.

T F **17.** Roundworms have a gut with only one opening.

T F **18.** The respiratory system permits the exchange of oxygen and carbon dioxide gas.

T F **19.** Flatworms have a well-developed cerebral ganglion, or brain, in one of their anterior segments.

T F **20.** Nerve nets are complex nervous systems.

T F **21.** Muscles are attached to the outside of exoskeletons.

T F **22.** The nervous system helps animals sense and respond to their environment.

• • • • • • • • • • • • • •
Complete each statement by writing the correct term or phrase in the space provided.

23. A(n) _____ is a collection of different cells that work together to perform a specific function.

24. Without a(n) _____ _____ , an animal could not eliminate the waste products of cellular metabolism.

25. The _____ _____ of an earthworm is formed from a fluid contained under pressure in a closed cavity.

26. A(n) _____ develops after a zygote undergoes cell division to form a hollow ball of cells.

27. In a bilaterally symmetrical animal, the top surface of the animal is referred to

as _____ and the bottom surface as _____ .

The front end of the animal is _____ and the back end is

_____ .

28. An animal has mobility not found in other multicellular organisms because

animal cells do not have rigid _____ _____ .

29. A(n) _____ _____ is a network of vessels that carry fluids to all parts of the body.

30. Muscles, most of the skeleton, the circulatory system, reproductive organs, and excretory organs arise from the primary tissue layer called

_____ .

31. A sea anemone's body plan is an example of _____

_____ because its body parts are arranged around a central axis.

32. Somites and the _____ _____ are examples of segmentation in vertebrates.

33. A terrestrial animal's _____ _____ helps the animal maintain its water balance.

34. An animal that is a(n) _____ has both ovaries and testes.

35. Ectoderm, endoderm, and mesoderm are called _____

_____ _____ because they give rise to all the tissues and organs of an adult body.

36. _____ are not suitable respiratory organs for most terrestrial animals because they must be kept moist.

37. The _____ _____ contributed to the evolution of complex organs composed of more than one tissue type.

• • • • • • • • • • • • • •
Read each question, and write your answer in the space provided.

38. Describe diploidy, and explain what advantage it provides to animals.

_____

_____

_____

_____

39. Describe the organization of the animal kingdom and at least six physical characteristics scientists consider to determine the evolutionary relationships between animals.

_____

_____

_____

_____

_____

_____

_____

40. Explain why animals that use external fertilization must release large numbers of gametes.

_____

_____

_____

_____

_____

41. List the three types of skeletal systems, and name two examples of each type.

_____

_____

_____

42. List the eight characteristics that all animals share.

_____

_____

_____

Question 43 refers to the figure below.

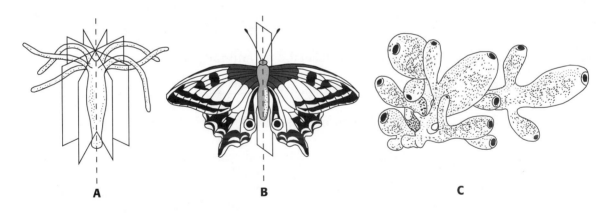

A   B   C

43. The figure above shows the different types of animal body symmetry. What type of body symmetry is labeled *A*? *B*? *C*?

_____

_____

_____

Name _____ Date _____ Class _____

Introduction to Animals

Use the terms from the list below to fill in the blanks in the following passage.

acoelomates	ectoderm	hydrostatic skeleton
asymmetrical	endoderm	internal fertilization
bilateral symmetry	endoskeleton	mesoderm
blastula	exoskeleton	open circulatory system
body plan	external fertilization	phylogenetic tree
cephalization	gastrovascular cavity	pseudocoelomates
closed circulatory system	gills	radial symmetry
coelom	hermaphrodites	respiration
coelomates		

In all animals except sponges, the zygote undergoes cell divisions that form

a(n) (1) _____ , which eventually develops into three distinct

layers of cells— (2) _____ , (3) _____ , and

(4) _____ .

All animals have their own particular (5) _____

_____ , a term used to describe an animal's shape,

symmetry, and internal organization. Sponges are (6) _____ .

The first animals to evolve in the ancient oceans had (7) _____

_____ , meaning the body parts are arranged around a central axis.
The bodies of all other animals have distinct right and left halves. This is called

(8) _____ _____ . Most animals with this
type of symmetry also have evolved an anterior concentration of sensory

structures and nerves—a process called (9) _____ .

Bilaterally symmetrical animals have different kinds of internal body plans

depending on whether they have a(n) (10) _____ , a body
cavity filled with fluid. Animals with no body cavity are called

(11) _____ . (12) _____ have a body cavity

located between the mesoderm and the endoderm. (13) _____
have a body cavity located entirely within the mesoderm. This means the gut
and other internal organs are suspended within a fluid-filled coelom.

To visually represent the relationships among various groups of animals, scientists often use a type of branching diagram called a(n)

(14) _____ _____ , which shows how animals are related through evolution.

The digestive system enables animals to ingest and digest food. Simple

animals have a(n) (15) _____ _____ , which has only one opening. More complex animals have a digestive tract (gut) with a mouth and an anus.

The uptake of oxygen and the release of carbon dioxide are called

(16) _____ and can take place only across a wet surface, such as the damp skin of an earthworm. In general, land animals use lungs

and aquatic animals use (17) _____ .

In complex animals, a system is needed to deliver oxygen and nutrients to the

cells. In a(n) (18) _____ _____

_____ the heart pumps a fluid into the body cavity and the fluid collects in open spaces in the animal's body and is returned to the heart. In a(n)

(19) _____ _____ _____ , the heart pumps blood through a system of blood vessels. The blood remains in the blood vessels and materials pass into and out of the blood vessels through diffusion.

An animal's skeleton provides a framework that supports its body and helps

protect its soft parts. Earthworms have a(n) (20) _____

_____ , which consists of water that is contained under pressure in a coelom. Insects, clams, and crabs have a(n)

(21) _____ , which is a hard, external skeleton that encases

the body of the animal. A(n) (22) _____ is composed of a hard material, such as bone, and is embedded within an animal.

In sexual reproduction, a new individual is formed by the union of a male and a female gamete. Many simple invertebrates, including slugs and earthworms, produce both types of gametes because they have both testes and ovaries.

Such animals are called (23) _____ . Most aquatic animals release the male and female gametes near one another in the water, where

fertilization occurs. This method is called (24) _____

_____ . In (25) _____ _____ , the union of the sperm and egg occurs within the female's body.

Introduction to Animals

Although animals have many features in common, there are also differences between specific groups of animals. The table below lists several features of six different animals. The table identifies similarities among some of the animals. Use the table to answer questions 1–6.

	Skeleton type	Mobility	Symmetry	Internal body plan	Segmentation
Eagle	Endoskeleton	Walk, fly	Bilateral	Coelomate	Present
Whale	Endoskeleton	Swim	Bilateral	Coelomate	Present
Goldfish	Endoskeleton	Swim	Bilateral	Coelomate	Present
Grasshopper	Exoskeleton	Walk, fly	Bilateral	Coelomate	Present
Sponge	Spicules, spongin	Sessile	Asymmetrical	Lack true tissues	Not present
Hydra	Hydrostatic skeleton	Glide, tumble	Radial	Gastrovascular cavity	Not present

Read each question, and write your answer in the space provided.

1. Which animals are not segmented?

_____

_____

_____

_____

2. How does the internal body plan of the animals you listed in question 1 differ from the internal body plan of the other animals?

_____

_____

_____

_____

3. Does the table indicate a relationship between segmentation and skeleton type? Explain.

_____

_____

_____

_____

4. Does the table indicate a relationship between symmetry and skeleton type? Explain.

_____

_____

_____

_____

5. What type of body symmetry do sponges and hydras have?

_____

_____

_____

_____

6. What conclusion can you draw about the relationship between internal body plan and segmentation?

_____

_____

_____

_____

Simple Invertebrates

In the space provided, write the letter of the term or phrase that best completes each statement or best answers each question.

_____ 1. Collar cells draw water through the sponge's many pores and into the internal cavity of the sponge by beating their
   a. osculum.
   b. cilia.
   c. flagella.
   d. tentacles.

_____ 2. Support for most sponges is provided by a simple skeleton composed of protein fibers called
   a. spongin.
   b. spicules.
   c. gemmules.
   d. silica.

_____ 3. Sessile animals
   a. live in marine environments.
   b. are made of cells.
   c. are attached to a fixed surface during their lives.
   d. have a large internal cavity.

_____ 4. Hydras attach themselves to rocks or water plants by means of a sticky secretion produced by the
   a. spicules.
   b. tegument.
   c. tentacles.
   d. basal disk.

_____ 5. The cnidarians known as jellyfish are members of the class
   a. Hydrozoa.
   b. Anthozoa.
   c. Scyphozoa.
   d. Cestoda.

_____ 6. Which of the following is characteristic of the roundworm *Ascaris?*
   a. The eggs can live in soil for years.
   b. They can block ducts leading from organs in the human body, such as the gallbladder.
   c. They can travel to the lungs and cause respiratory distress.
   d. all of the above

_____ 7. Tapeworms absorb food from the host's intestine through their
   a. skin.
   b. mouth.
   c. digestive system.
   d. collar cells.

_____ 8. The simplest animal that has a one-way digestive system is the
   a. roundworm.
   b. flatworm.
   c. fluke.
   d. trematode.

_____ 9. The body form called a medusa is
   a. usually found with the tentacles pointing downward.
   b. usually found with the tentacles pointing upward.
   c. common to sponges, anemones, and corals.
   d. found only in sessile organisms.

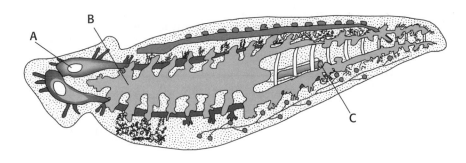

_____ 10. The structure labeled *A* is
    a. the brain.   **c.** the mouth.
    b. a nerve cord.   **d.** an eyespot.

_____ 11. The structure labeled *B* is
    a. the intestine.   **c.** the mouth.
    b. a flame cell.   **d.** an eyespot.

_____ 12. The structure labeled *C* is
    a. the intestine.   **c.** the mouth.
    b. a flame cell.   **d.** the anus.

• • • • • • • • • • • • • • •

In the space provided, write the letter of the description that best matches the term or phrase.

_____ 13. amoebocyte

_____ 14. spongin

_____ 15. spicules

_____ 16. gemmules

_____ 17. Hydrozoa

_____ 18. planulae

_____ 19. Scyphozoa

_____ 20. Anthozoa

_____ 21. flukes

_____ 22. endoparasites

_____ 23. ectoparasites

_____ 24. flame cells

_____ 25. tegument

_____ 26. proglottids

a. in planarians, specialized cells with beating tufts of cilia that draw water through pores to the outside of the worm's body

b. tough, flexible protein fibers that make up the skeleton of most sponges

c. the rectangular body sections of the tapeworm

d. free-swimming cnidarian larvae

e. an amoebalike cell in a sponge that moves through the body cells, supplying nutrients and removing wastes

f. thick protective covering of cells that protects endoparasites from being digested by their host

g. class that includes the most primitive cnidarians

h. tiny needles of silica or calcium carbonate that make up the skeletons of some sponges

i. clusters of amoebocytes with protective coats that enable them to survive harsh conditions that may kill the adult sponge; produced by some freshwater sponges

j. parasites that live on the external surfaces of their hosts

k. the largest class of cnidarians, which includes sea anemones and corals

l. class of cnidarians that includes jellyfish

m. parasites that live inside their hosts

n. parasitic flatworms of the class Trematoda

Complete each statement by writing the correct term or phrase in the space provided.

27. An organism that produces both eggs and sperm is called a(n)

_____ .

28. Some sponges have both _____ and _____ to support and strengthen their bodies.

29. In most sponge species, the eggs of an individual sponge cannot be fertilized by

the sperm of the _____ _____ .

30. The Portuguese man-of-war is a member of the class _____ .

31. Most hydrozoans are colonial organisms whose polyps reproduce

_____ by forming small buds on the outside of their bodies.

32. Anthozoans typically have a stalklike body topped by a crown of

_____ .

33. *Dugesia* must extend its muscular _____ out of its centrally located mouth in order to feed.

34. Humans can become infected by beef tapeworms if they eat beef that is not

_____ .

35. *Schistosoma*, sometimes called a blood fluke, must live in a(n)

_____ before it can infect humans.

36. *Trichinella*, which can be found in undercooked pork, causes a disease called

_____ in humans.

Read each question, and write your answer in the space provided.

37. Describe how sponge cells get nutrients.

_____

_____

_____

_____

_____

38. Describe the process of sexual reproduction in sponges.

_____

_____

_____

39. What type of specialization is found among the polyps that make up a Portuguese man-of-war?

_____

_____

_____

40. How do most hydrozoans reproduce? How do scyphozoans reproduce?

_____

_____

_____

41. *Obelia* produces both polyps and gametes that join to produce planulae. Explain what the life of each form will be like.

_____

_____

_____

42. How do planarians reproduce?

_____

_____

_____

43. How do tapeworms obtain nutrients without a mouth or a digestive system?

_____

_____

_____

44. How can humans prevent hookworm infestation?

_____

_____

_____

Simple Invertebrates

In the space provided, write the letter of the description that best matches the term or phrase.

_____ 1. mesoglea

_____ 2. ostia

_____ 3. sessile

_____ 4. oscula

_____ 5. choanocytes

_____ 6. amoebocytes

_____ 7. spongin

_____ 8. spicules

_____ 9. gemmules

_____ 10. medusa

_____ 11. polyp

_____ 12. cnidocytes

_____ 13. nematocyst

_____ 14. basal disk

_____ 15. planula

_____ 16. proglottids

_____ 17. fluke

_____ 18. tegument

a. sponge cells that have irregular amoebalike shapes

b. clusters of amoebocytes encased in protective coats

c. large openings in a sponge's body wall

d. resilient flexible protein fiber

e. free-floating life–form of a cnidarian

f. stinging cells located on tentacles of cnidarians

g. firmly attached to the sea bottom or other surface

h. body form of a cnidarian, which is attached to a rock or some other object

i. a gel-like substance

j. small barbed harpoon inside a cnidocyte

k. flagellated cells also known as collar cells

l. larval stage of a hydrozoan

m. body sections of flatworms

n. parasitic flatworm

o. tiny needles of silica or calcium carbonate that form a sponge's skeleton

p. protective covering of endoparasitic flukes

q. area on *Hydra* that produces a sticky secretion

r. tiny openings or pores in a sponge's body wall

CHAPTER

30 **TEST PREP PRETEST**

Mollusks and Annelids

In the space provided, write the letter of the term or phrase that best completes each statement or best answers each question.

_____ 1. The fertilized eggs of both mollusks and annelids develop into a distinct larval form called a
  a. polyp.
  b. veliger.
  c. trochophore.
  d. nudibranch.

_____ 2. Most mollusks have a
  a. three-chambered heart and a closed circulatory system.
  b. three-chambered heart and an open circulatory system.
  c. four-chambered heart and an open circulatory system.
  d. four-chambered heart and a closed circulatory system.

_____ 3. Which of the following is NOT a characteristic of mollusks?
  a. acoelomate body structure
  b. bilateral symmetry
  c. organ systems
  d. three-part body plan

_____ 4. Annelids were the first organisms to exhibit
  a. a true coelom.
  b. organ systems.
  c. bilateral symmetry.
  d. segmentation.

_____ 5. All annelids have a(n)
  a. closed circulatory system and a three-chambered heart.
  b. closed circulatory system and a four-chambered heart.
  c. closed circulatory system and a series of hearts.
  d. open circulatory system and a series of hearts.

_____ 6. Which of the following is NOT a characteristic of annelids?
  a. gills or lungs
  b. complex organ systems
  c. a highly specialized gut
  d. segmented bodies

_____ 7. When soil in the digestive tract of an earthworm leaves the crop, it passes to the
  a. pharynx.
  b. gizzard.
  c. esophagus.
  d. anus.

_____ 8. The movement of earthworms requires
  a. muscles lining the interior body wall.
  b. muscle contractions.
  c. traction provided by setae.
  d. All of the above

Circle T *if the statement is true or* F *if it is false.*

T F **9.** Mollusks and annelids are NOT related.

T F **10.** Most mollusks breathe with ciliated gills located in their mantle cavity.

T F **11.** Octopuses and squids have a closed circulatory system.

T F **12.** Bivalves reproduce asexually.

T F **13.** During evolution, slugs and nudibranchs lost their shells.

T F **14.** The species of the class Cephalopoda are sessile filter feeders.

T F **15.** Cephalopods are the most intelligent of all invertebrates.

T F **16.** Some annelids have fleshy appendages called parapodia.

T F **17.** The muscular regions that provide movement for a mollusk are called parapodia.

T F **18.** Polychaetes are terrestrial worms that tunnel through the soil.

T F **19.** Earthworms are hermaphrodites.

Complete each statement by writing the correct term or phrase in the space provided.

20. Mollusks and annelids were probably the first major groups to develop a true

_____ .

21. Mollusks filter useful materials from their coelomic fluid by tiny tubular

structures called _____ .

22. When the _____ muscles of a bivalve contract, they cause the valves to close forcefully.

Questions 23–25 refer to the figure below, which shows the structure of an earthworm.

23. The structure labeled *A*, called the _____ , grinds up soil that the earthworm ingests.

24. The _____ _____ , labeled *B*, coordinates the muscular activity of each body segment.

25. The earthworm anchors several of its segments by sinking its

_____ , labeled *C*, into the ground.

26. Bivalves feed by sucking sea water through hollow tubes called

_____ .

27. All mollusks EXCEPT bivalves have a rasping tonguelike organ called a(n)

_____ .

28. _____ have a primitive brain called a cerebral ganglion located in one anterior segment.

29. Leeches differ from other annelids in that they lack both _____

and _____ .

• • • • • • • • • • • • • •
Read each question, and write your answer in the space provided.

30. What is a trochophore?

_____

_____

_____

31. Why are terrestrial snails more active when the air around them is moist?

_____

_____

_____

32. Describe the function of septa in annelids.

_____

_____

_____

33. How are cephalopods adapted as predators?

_____

_____

_____

34. In what basic way do the annelid and mollusk body plans differ?

_____

_____

_____

_____

35. How are annelids classified?

_____

_____

_____

_____

36. Why do earthworms require a moist environment?

_____

_____

_____

_____

Mollusks and Annelids

*Use the terms from the list below to fill in the blanks in the
following passage.*

adductor muscles	nephridia	setae
cerebral ganglion	parapodia	siphons
foot	radula	trochophore
mantle	septa	visceral mass

Mollusks and the annelids were probably the first major groups of

organisms to develop a true coelom. Another feature shared by many mollusks

and annelids is a larval stage called a(n) (1) _____ , which

develops from the fertilized egg.

The body cavity in mollusks is a true coelom and usually exhibits

bilateral symmetry. Mollusks have many organ systems, which are contained

in the (2) _____ _____ .

A(n) (3) _____ wraps around the visceral mass.

Every mollusk has a muscular region called a(n) (4) _____ .

Many mollusks have one or two shells, which protect their soft bodies. All

mollusks, except bivalves, have a rasping tonguelike organ called a(n)

(5) _____ .

Mollusks are the only coelomates without segmented bodies. Like

roundworms, mollusks have a one-way digestive system. Mollusks use their

coelom as a collecting place for waste-laden body fluids. Before leaving

the body, this fluid passes into tiny tubular structures called

(6) _____ . Mollusks have a circulatory system and respire

through gills.

(continued on next page)

Gastropods—snails and slugs—are primarily a marine group that have also very successfully invaded fresh water and terrestrial habitats. Most bivalves are marine, but some live in fresh water. All bivalves have a two-part hinged shell. Two thick (7) _____ _____ connect the valves, and when contracted they cause the valves to close tightly.

Most bivalves are filter feeders, and many use their muscular foot to dig down into the sand. Once there, the cilia on their gills draw in sea water through hollow tubes called (8) _____ .

Annelids are easily recognized by their segments, which are visible externally as a series of ringlike structures along the length of their body. A well-developed primitive brain, called a(n) (9) _____ _____ , is located in one anterior segment. Internal body walls, called (10) _____ , separate the segments of most annelids.

Annelids have a body cavity that is a true coelom, and they have organ systems. Most annelids have external bristles called (11) _____ . Some annelids have fleshy appendages called (12) _____ . These two external characteristics are used to classify annelids.

CHAPTER
31 **TEST PREP PRETEST**

Arthropods

In the space provided, write the letter of the term or phrase that best completes each statement or best answers each question.

_____ 1. Subphylum Uniramia includes
   a. insects.
   b. millipedes.
   c. centipedes.
   d. All of the above

_____ 2. Living arthropods are traditionally separated into two large groups based on their
   a. mouthparts.
   b. legs.
   c. antennae.
   d. eyes.

_____ 3. Arthropods periodically shed and discard their exoskeletons as they grow in a process called
   a. reproduction.
   b. segmentation.
   c. ecdysis.
   d. metamorphosis.

_____ 4. Members of the arthropod subphylum Chelicerata include
   a. crabs and lobsters.
   b. spiders and scorpions.
   c. insects.
   d. millipedes and centipedes.

_____ 5. All arachnids, except some mites, are
   a. insectivores.
   b. herbivores.
   c. carnivores.
   d. omnivores.

_____ 6. Most crustaceans have a distinctive larval form called a(n)
   a. egg.
   b. juvenile.
   c. pupa.
   d. nauplius.

_____ 7. Centipedes have a head region followed by up to about
   a. 15 segments.
   b. 30 segments.
   c. 170 segments.
   d. 1,000 segments.

_____ 8. Which of the following characteristics is NOT shared by all insects?
   a. three body sections
   b. five-part radial symmetry
   c. three pairs of legs
   d. one pair of antennae

_____ 9. In a young insect's final molt to the adult stage, it undergoes a process of physical change called
   a. segmentation.
   b. reproduction.
   c. metamorphosis.
   d. ecdysis.

_____ 10. The head, thorax, and abdomen of mites
   a. are separate segmented sections.
   b. form two sections, the cephalothorax and the abdomen.
   c. are fused to form a single body.
   d. form two sections, the head and a fused thorax and abdomen.

_____ 11. Insect wings are attached to the
- **a.** head.
- **b.** appendages.
- **c.** abdomen.
- **d.** thorax.

_____ 12. Spiders produce silk from
- **a.** spinnerets.
- **b.** mandibles.
- **c.** chelicerae.
- **d.** pedipalps.

_____ 13. Specialized structures in arthropods, called Malpighian tubules, are important in
- **a.** respiration.
- **b.** excretion.
- **c.** reproduction.
- **d.** sensing movement.

_____ 14. The three main body segments of all arthropods are the
- **a.** cephalothorax, abdomen, and appendages.
- **b.** head, cephalothorax, and appendages.
- **c.** thorax, appendages, and spiracles.
- **d.** head, thorax, and abdomen.

• • • • • • • • • • • • • •

In the space provided, write the letter of the description that best matches the term or phrase.

_____ 15. nauplius

_____ 16. Chelicerata

_____ 17. segmentation

_____ 18. cephalothorax

_____ 19. exoskeleton

_____ 20. spinnerets

_____ 21. compound eye

_____ 22. tracheae

_____ 23. spiracle

_____ 24. Malpighian tubules

_____ 25. pedipalps

_____ 26. carapace

_____ 27. swimmerets

_____ 28. uropods

_____ 29. telson

_____ 30. Isopoda

- **a.** flattened, paddlelike appendages at the end of a decapod's abdomen
- **b.** a body section of some arthropods formed by the fusion of the head with the thorax
- **c.** larval form of crustaceans
- **d.** a decapod's tail spine
- **e.** a network of fine tubes used for respiration by many arthropods
- **f.** the shell-like structure that encases the bodies of arthropods
- **g.** a terrestrial order of crustaceans that includes pill bugs and sow bugs
- **h.** subphylum that includes arthropods with fangs
- **i.** a second pair of appendages used by arachnids to catch and handle prey
- **j.** appendages attached along the underside of the abdomen that are used by lobsters and crayfish for swimming and reproduction
- **k.** the term for having a body divided into sections
- **l.** the shield that covers a decapod's cephalothorax
- **m.** fingerlike excretory organs
- **n.** modified appendages at the end of the abdomen for producing silk and adhesive
- **o.** an opening that functions during respiration in many arthropods
- **p.** made of thousands of individual units, each with its own lens and retina

Complete each statement by writing the correct term or phrase in the space provided.

31. The _____ , early arthropods, were the first animals with eyes capable of forming images.

32. _____ were the first animals with wings to evolve.

33. An arthropod must shed its _____ to grow.

34. In the grasshopper, three fused ganglia serve as the _____ .

35. In incomplete metamorphosis, the juvenile, or _____ , is essentially a smaller version of the adult.

36. The role played by an individual ant in a colony is called its _____ .

37. In the grasshopper, there are four kinds of mouthparts: labium and labrum,

which function as _____ ; mandibles, which function as

_____ ; and maxillas, which function as _____ .

38. A compound eye sees _____ more quickly than the vertebrate eye but does not see as clearly.

39. Deer ticks may carry the virus that causes _____

_____ .

40. The free-swimming nauplius is the _____ form of a crustacean.

Questions 41–43 refer to the figure below.

A   B   C   D

41. The process taking place in the figure above is _____

_____ .

42. The stage labeled *D* shows the _____ , while the stage labeled

A shows the _____ .

43. During this process, the _____ is closed within a protective

capsule called a(n) _____ , labeled *C*.

44. Crabs and lobsters are members of the subphylum of arthropods known as

_____ .

45. Arachnids have _____ pairs of legs.

46. _____ are the only arthropods with two pairs of antennae.

47. In terms of diet, centipedes are _____ , while millipedes are

_____ .

• • • • • • • • • • • • • • •
Read each question, and write your answer in the space provided.

48. Why do scientists think that the ancestors of arthropods are annelids?

_____

_____

49. List four important characteristics of arthropods.

_____

_____

50. List three important characteristics of crustaceans.

_____

_____

51. Distinguish between centipedes and millipedes.

_____

_____

_____

52. Describe the wings of grasshoppers.

_____

_____

53. What is the evolutionary advantage of complete metamorphosis in insects?

_____

_____

_____

54. List three ways in which the mouthparts of insects might be adapted to their diet.

_____

_____

CHAPTER
31 **VOCABULARY**

Arthropods

Write the correct term from the list below in the space next to its definition.

appendages   Malpighian tubules   spinneret   thorax
cephalothorax   mandible   spiracle   tracheae
chelicerae   pedipalps

_____ **1.** structures that extend from the body wall

_____ **2.** the midbody region

_____ **3.** head fused with thorax

_____ **4.** network of tubes through which arthropods respire

_____ **5.** structure through which air from outside of the arthropod's body passes

_____ **6.** excretory system of arthropods

_____ **7.** mouthparts of the subphylum Chelicerata

_____ **8.** pairs of appendages modified to handle prey

_____ **9.** secretes strands of silk

_____ **10.** chewing mouthpart of subphylum Uniramia

In the space provided, write the letter of the description that best matches the term or phrase.

_____ **11.** molting

_____ **12.** metamorphosis

_____ **13.** chrysalis

_____ **14.** pupa

_____ **15.** nymph

_____ **16.** caste

_____ **17.** nauplius

_____ **18.** compound eye

_____ **19.** krill

a. the role played by an individual in a colony

b. the physical change of a young insect into an adult

c. a young insect that looks like a small wingless adult

d. larval form of crustacean

e. stage in metamorphosis during which a young insect becomes an adult

f. a protective capsule

g. periodic shedding of exoskeleton

h. small marine crustacean

i. made of thousands of individual units

Echinoderms and Invertebrate Chordates

In the space provided, write the letter of the term or phrase that best completes each statement or best answers each question.

_____ 1. An animal whose mouth develops from the blastopore is called a
  a. deuterostome.   **c.** cephalostome.
  b. protostome.   **d.** pseudocoelom.

_____ 2. Echinoderms share all of the following characteristics EXCEPT
  a. an endoskeleton composed of ossicles.
  b. a radially symmetrical body plan in adulthood.
  c. a water-vascular system.
  d. a notochord.

_____ 3. Animals that have their body parts arranged around a central point are said to be
  a. asymmetric.   **c.** bilaterally symmetrical.
  b. radially symmetrical.   **d.** spherically symmetrical.

_____ 4. Deuterostomes include all of the following EXCEPT
  a. vertebrates.   **c.** arthropods.
  b. sea stars.   **d.** tunicates.

_____ 5. The water-vascular system of echinoderms functions as a
  a. means of movement.
  b. gas exchange system.
  c. waste excretion system.
  d. All of the above

_____ 6. Sea lilies and feather stars are unlike all other living echinoderms because their mouth
  a. is located on the lower surface.
  b. develops from the blastopore.
  c. is disc shaped.
  d. is located on the upper surface.

_____ 7. Chordates are characterized by all of the following EXCEPT
  a. radial symmetry.
  b. pharyngeal slits.
  c. a tail that extends beyond the anus.
  d. a dorsal, hollow nerve cord.

_____ 8. The endoskeleton of chordates allows them to
  a. swim through water.
  b. grow large and move quickly.
  c. swing their bodies from side to side.
  d. All of the above

_____ 9. The adult tunicate develops a tough sac around its body called a(n)
  a. tunic.   **c.** endoskeleton.
  b. scale.   **d.** ossicle.

_____ **10.** Tunicates and lancelets are similar in that both
  a. are sessile.
  b. are covered with a tunic.
  c. are invertebrate chordates.
  d. have backbones.

_____ **11.** Which of the following invertebrate chordates is sessile?
  a. sea star   **c.** lancelet
  b. tunicate   **d.** hydra

• • • • • • • • • • • • • • •
Circle T *if the statement is true or* F *if it is false.*

T F **12.** In deuterostomes, the mouth develops from or near the blastopore.

T F **13.** Brittle stars, sea cucumbers, and chordates share the same pattern of early embryonic development.

T F **14.** Echinoderms were the first animals to develop an endoskeleton.

T F **15.** The ossicles of the adult echinoderm fuse to become an exoskeleton.

T F **16.** The coelomic circulation of sea stars enables them to move across the sea floor using tube feet.

T F **17.** The body cavity of the echinoderm functions as a simple circulatory and respiratory system.

T F **18.** All chordates are deuterostomes.

T F **19.** Echinoderms have openings called pharyngeal slits in their pharynx.

T F **20.** Members of the subphyla Urochordata and Cephalochordata are invertebrate chordates.

T F **21.** The endoskeleton of some chordates is completely internal.

T F **22.** Both tunicates and lancelets are hermaphrodites as adults.

• • • • • • • • • • • • • • •
Complete each statement by writing the correct term or phrase in the space provided.

23. In _____ , the mouth not only forms later than the

_____ but also forms on another part of the embryo.

24. Protostomes include coelomate animals, such as _____ ,

_____ , and _____ .

25. Many echinoderms crawl across the seafloor by means of a(n)

_____ _____ system.

26. Because echinoderms and chordates develop similarly as embryos, it is likely

that they are derived from a(n) _____ _____ .

27. In many echinoderms, respiration and waste removal are aided by

_____ _____ , which are small fingerlike

projections that grow between the spines.

28. The echinoderm endoskeleton is composed of individual plates called

_____ .

29. The water-vascular system of the echinoderm is a series of interconnected

_____ and thousands of tiny hollow _____

_____ .

30. The water-vascular system helps an echinoderm crawl and is also important for

_____ and _____ _____ .

31. Echinoderms have no head or brain, but they do have a(n) _____

of _____ with branches that extend into each of the arms.

32. Many species of sea stars have ossicles that produce pincerlike structures called

_____ .

33. _____ _____ differ from other echinoderms in that their ossicles are small and not connected.

34. Of all the chordate characteristics, only the _____

_____ are retained by the adult tunicate.

35. During the development of the chordate embryo, a stiff rod called a(n)

_____ develops along the back.

36. The echindoderm's coelom functions as a(n) _____ and

_____ system.

37. Some tunicates reproduce asexually by _____ .

• • • • • • • • • • • • • • • • • •

Read each question, and write your answer in the space provided.

Questions 38–40 refer to the figure at right.

38. Identify the animal shown at right.

_____

39. Based on the labeled structures, how would you know that this animal is a chordate?

_____

_____

Notochord
Dorsal nerve cord
Pharynx
Intestine
Segmented muscles
Tail

40. How would an adult tunicate differ from the animal shown on page 127?

_____

41. Describe how a sea star is able to move using its water-vascular system.

_____

_____

_____

_____

_____

_____

42. Describe the similarities between tunicates and lancelets.

_____

_____

_____

Questions 43 and 44 refer to the figure below, which shows the gastrula stage of a protostome embryo and a deuterostome embryo.

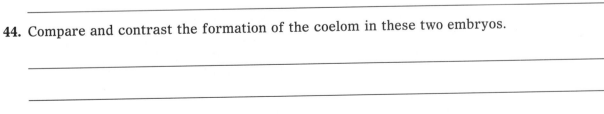

43. Compare the location of the gut in these two embryos.

_____

_____

44. Compare and contrast the formation of the coelom in these two embryos.

_____

_____

_____

CHAPTER
32 **VOCABULARY**

Echinoderms and Invertebrate Chordates

Write the correct term from the list below in the space next to its definition.

blastopore   ossicles   skin gills
deuterostome   protostome   water-vascular system

_____ 1. the opening to the outside during the gastrula stage of an embryo

_____ 2. an animal whose mouth develops from or near the blastopore

_____ 3. an animal whose anus develops from or near the blastopore

_____ 4. calcium-rich plates that make up the endoskeleton of an echinoderm

_____ 5. a water-filled system of interconnected canals and thousands of tiny, hollow-tube feet

_____ 6. small fingerlike projections that grow between the spines of an echinoderm where respiratory gases are exchanged.

In the space provided, write the letter of the description that best matches the term or phrase.

_____ 7. chordate

_____ 8. notochord

_____ 9. pharyngeal slit

_____ 10. invertebrate chordate

a. openings that develop in the wall of the pharynx

b. chordates that do not have backbones

c. animals that have a notochord

d. a stiff rod that develops along the back of an embryo

Name_____ Date _____ Class_____

Introduction to Vertebrates

In the space provided, write the letter of the term or phrase that best completes each statement or best answers each question.

_____ 1. For about 100 million years, the oceans were dominated by
     a. amphibians.   **c.** jawless fishes.
     b. reptiles.   **d.** mammals.

_____ 2. Which of the following is NOT a key adaptation that enabled fish to dominate the oceans?
     a. paired fins   **c.** streamlined bodies
     b. lungs   **d.** strong jaws

_____ 3. The first vertebrates to live on land were
     a. amphibians.   **c.** fishes.
     b. reptiles.   **d.** mammals.

_____ 4. Amphibians successfully invaded land because they evolved
     a. legs and lungs.
     b. lungs and watertight eggs.
     c. a backbone and a chambered heart.
     d. solid bones.

_____ 5. The pattern of bones in an amphibian's limbs resembles the pattern in a
     a. lobe-finned fish.   **c.** jawless fish.
     b. cartilaginous fish.   **d.** shark.

_____ 6. One adaptation that helped reptiles become the first fully terrestrial vertebrates was
     a. bony scales.
     b. legs.
     c. lungs.
     d. skin that was almost watertight.

_____ 7. The ability of dinosaurs to be fast, agile runners was due to
     a. a tail that could be used for balance.
     b. a strong calcium carbonate skeleton.
     c. their large size.
     d. legs positioned directly under the body.

_____ 8. A carnivorous dinosaur that evolved by the late Jurassic period and had sickle-shaped claws on its feet was the
     a. sauropod.   **c.** thecodont.
     b. theropod.   **d.** placoderm.

_____ 9. Which of the following describes an animal whose metabolism is too slow to produce enough heat to warm its body?
     a. ectothermic   **c.** warm-blooded
     b. endothermic   **d.** none of the above

10. The disappearance of dinosaurs 65 million years ago was most likely caused by
   a. an earthquake.
   b. a gigantic meteorite or comet colliding with Earth.
   c. a solar eclipse.
   d. rising sea levels.

11. After the Cretaceous extinction, the only animals that survived were
   a. animals that could live in the oceans.
   b. animal species smaller than a small dog.
   c. the theropods.
   d. None of the above

12. The only mammals that lay eggs are
   a. monotremes.
   b. placentals.
   c. marsupials.
   d. birds.

13. Features that at first led scientists to classify *Archaeopteryx* as a dinosaur include
   a. a breastbone.
   b. feathers.
   c. solid bones and teeth.
   d. a fused collarbone.

Questions 14–16 refer to the figure below.

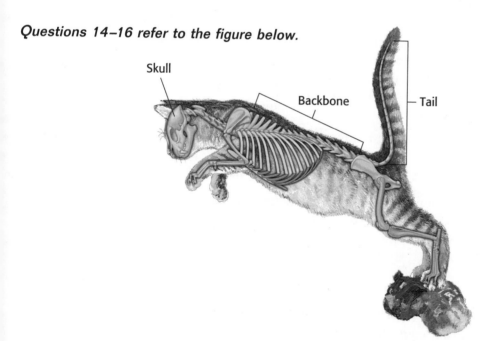

Skull

Backbone

Tail

14. The kind of chordate represented by this skeleton is a(n)
   a. tunicate.
   b. lancelet.
   c. vertebrate.
   d. None of the above

15. The individual segments that make up the backbone are called
   a. scales.
   b. organs.
   c. bony plates.
   d. vertebrae.

16. An advantage to having a backbone is that the backbone
   a. provides support for the dorsal nerve cord.
   b. provides a site for muscle attachment.
   c. protects the dorsal nerve cord.
   d. All of the above

T   F   **17.** In most vertebrates, the notochord is replaced by the backbone.

T   F   **18.** In vertebrates, support and protection of the dorsal nerve cord are provided by strong muscle attachments.

T   F   **19.** The earliest fishes had strong jaws and cartilaginous bodies.

T   F   **20.** Early fishes became flattened sideways, making movement through the water easier.

T   F   **21.** Cartilage is a very heavy, strong, and flexible tissue.

T   F   **22.** The first cartilaginous fishes and the bony fishes evolved at about the same time.

T   F   **23.** The lungs of amphibians evolved for gas exchange.

T   F   **24.** Agnathans, the earliest fishes, lacked jaws and paired fins, and they are represented today by the lampreys.

T   F   **25.** Dinosaurs dominated the land for about 150 million years.

T   F   **26.** Today, the largest group of vertebrates is birds.

T   F   **27.** Most vertebrates have a bony skull, a true coelem, radial symmetry, and complex sense organs.

Complete each statement by writing the correct term or phrase in the space provided.

28. All vertebrates have a(n) _____ circulatory system with a(n) _____ heart.

29. The organization of organs into _____ _____ is characteristic of all vertebrates.

30. The development of strong _____ , as in the acanthodians, was a key evolutionary innovation.

31. By the end of the _____ period, almost all of the early fishes had disappeared.

32. The three major groups of fishes living today are the _____ , the _____ fishes, and the _____ fishes.

33. A shark's skeleton is made of _____ .

34. _____ animals live their entire life cycle on land.

35. Mammals likely evolved from _____ , early reptiles that were probably _____ .

36. _____ mammals are nourished by the mother through an organ called the _____ .

37. List the order in which the major groups of vertebrates evolved.

_____

_____

38. What two important challenges did fishes have to meet in order to survive as predators in the water? What adaptations enabled them to meet these challenges?

_____

_____

_____

_____

_____

39. Contrast the environments inhabited by amphibians during the Age of Amphibians and during the Permian period.

_____

_____

_____

_____

_____

40. In what way are amphibians not fully adapted to life on land?

_____

_____

41. Why were reptiles able to become the dominant animals on Earth?

_____

_____

_____

42. Describe the debate over whether dinosaurs were ectothermic or endothermic.

_____

_____

_____

_____

CHAPTER 33 **VOCABULARY**

Introduction to Vertebrates

Complete the crossword puzzle using the clues provided.

ACROSS

1. maintain a high, constant body temperature from heat produced by metabolism
3. extinct, spiny fish
5. entire life cycle lived on land
7. a single supercontinent that existed 200 million years ago
8. extinct crocodilelike dinosaur

DOWN

1. body temperature is determined by the temperature of the environment
2. lightweight, strong, and flexible tissue
3. extinct, jawless fish
4. individual segment of a backbone
6. extinct order of reptiles

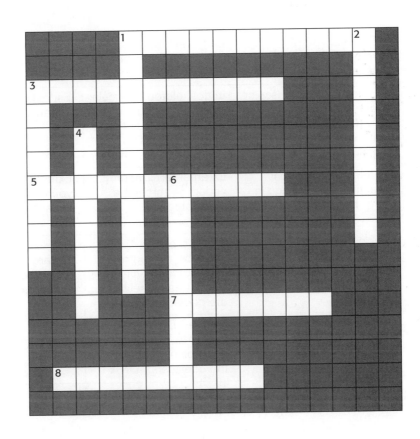

CHAPTER
33 SCIENCE SKILLS: COMPARING STRUCTURES/INTERPRETING TABLES

Introduction to Vertebrates

Amphibians, which include frogs, toads, and salamanders, were the earliest tetrapods, or four-legged land animals. They first appear in the fossil record about 360 million years ago. Scientists have long agreed that amphibians arose from one of three groups of lobe-finned fishes, whose fossils date to nearly 400 million years ago. These three groups are the lungfish and coelacanths, both of which survive today, and the (extinct) rhipidistians.

The table below lists some of the physical characteristics scientists have used to determine the closest living relative of amphibians. Use the information above and in the table below to answer questions 1–3 on page 227.

	Coelacanth	Lungfish	Amphibians
Digestive system	One-way	One-way	One-way
Respiratory system	Gills and primitive lungs; (fossils of rhipidistian relatives had internal nostrils)	Gills and primitive lungs; internal nostrils very similar to those of amphibians	Gills in young; adults breathe through lungs and skin; internal nostrils
Circulatory system	Early version of vertebrate chambered heart	Early version of vertebrate chambered heart	Primitive chambered heart
Skeletal system	Endoskeleton; muscular lobes in fins supported by jointed bones that are similar to limb bones in land animals	Endoskeleton; muscular lobes in fins supported by jointed bones that are similar to limb bones in land animals.	Endoskeleton; jointed appendages
Reproduction	Internal fertilization; female retains eggs in her body; gives birth to live young	External fertilization; young go through a larval stage	External fertilization; young go through a larval stage
Fossil record	Showed bones in fins could be related to those of land animals	Showed bones in fins could be related to those of land animals, but not as closely as the coelacanth's	Showed bone structure of early amphibians resembles that of lobe-finned fishes
DNA analysis		Showed DNA more similar to that of amphibians than was the coelacanth DNA	

Read each question, and write your answer in the space provided.

1. What evidence favors the coelacanth as the closest living relative of amphibians?

_____

_____

_____

_____

_____

2. Assume you had to convince someone that the lungfish was the closest relative of amphibians. Which physical characteristics of these animals would you discuss?

_____

_____

_____

_____

_____

3. DNA analysis is considered to be the most accurate evidence of the evolutionary relationships of animals. What does the DNA evidence indicate about the evolutionary relationship between lungfish and amphibians and coelacanths and amphibians?

_____

_____

_____

_____

_____

CHAPTER

34 **TEST PREP PRETEST**

Fishes and Amphibians

In the space provided, write the letter of the term or phrase that best completes each statement or best answers each question.

_____ 1. The major respiratory organ of a fish is the
   a. swim bladder.   **c.** gill.
   b. lung.   **d.** operculum.

_____ 2. Depending on the species, fish can reproduce through
   a. internal fertilization.   **c.** conjugation.
   b. spawning.   **d.** Both (a) and (b)

_____ 3. Lampreys and hagfishes are the only remaining
   a. bony fishes.   **c.** Osteichthyes.
   b. Chondrichthyes.   **d.** agnathans.

_____ 4. Because the bodies of freshwater fish contain more ions than the surrounding water, they
   a. excrete concentrated urine.
   b. need to take in salts from the environment.
   c. tend to take in water through osmosis.
   d. Both (b) and (c)

_____ 5. Members of the order Apoda do NOT
   a. use cutaneous respiration.   **c.** have legs.
   b. lay eggs.   **d.** bear live young.

_____ 6. Which of the following characteristics is NOT shared by all fishes?
   a. gills   **c.** single-loop circulation
   b. vertebral column   **d.** scales

_____ 7. In order to rise in the water, bony fishes fill their swim bladder with gas from
   a. their gills.   **c.** the surrounding water.
   b. their bloodstream.   **d.** their lungs.

_____ 8. Fish "swallow" water to
   a. force it from the mouth and over the gills.
   b. steady themselves in the water.
   c. feed.
   d. All of the above

_____ 9. Compared with that of a fish, a frog's cerebrum is
   a. less complex.   **c.** smaller.
   b. more complex.   **d.** None of the above

_____ 10. Jawless fishes have
   a. unpaired fins.   **c.** operculums.
   b. scales.   **d.** swim bladders.

_____ **11.** A modern amphibian's heart has a

    a. divided atrium and one ventricle.
    b. divided ventricle and one atrium.
    c. septum.
    d. Both (a) and (c)

_____ **12.** Frogs and toads are members of the order

    a. Anura.   **c.** Urodela.
    b. Apoda.   **d.** Caecilia.

_____ **13.** Because of their tympanic membrane, leopard frogs can hear well in

    a. both water and air.   **c.** air but not in water.
    b. water but not in air.   **d.** None of the above

Questions 14 and 15 refer to the figure at right, which shows the structure of a fish.

_____ **14.** The structure labeled *A* is called the

    a. dorsal fin.
    b. brain.
    c. operculum.
    d. lateral line.

_____ **15.** The structure labeled *B* is called the

    a. pectoral fin.   **c.** pelvic fin.
    b. operculum.   **d.** jaw.

Circle T *if the statement is true or* F *if it is false.*

T   F   **16.** The first fishes had no fins.

T   F   **17.** In fish, blood collects in the sinus venosus and passes into the atrium.

T   F   **18.** Marine fish excrete concentrated urine to prevent excessive water loss.

T   F   **19.** The only remaining agnathans are bony fishes.

T   F   **20.** Skates and rays have streamlined bodies.

T   F   **21.** All members of the order Urodela retain gills as adults.

T   F   **22.** Fishes have a double-loop circulation system.

T   F   **23.** Most bony fishes have a hard plate, called the operculum, that covers the gills on each side of the head.

T   F   **24.** The eggs of yellow perch are fertilized internally.

T   F   **25.** Spawning is a method of external fertilization.

T   F   **26.** The gills of bony fishes are more efficient than amphibians' lungs.

T   F   **27.** All modern amphibians use only their lungs for respiration.

T   F   **28.** All frogs, toads, salamanders, and caecilians have legs that enable them to move efficiently in terrestrial habitats.

T   F   **29.** The leopard frog protects itself from predators by means of skin glands that secrete poisonous substances.

Complete each statement by writing the correct term or phrase in the space provided.

30. The first fishes breathed using _____ .

31. By moving certain _____ and the _____ , bony fishes can move water over their gills while remaining stationary.

32. In amphibians, oxygen-rich blood and oxygen-poor blood mix in the

    _____ because it lacks a(n) _____ .

33. The only jawless fishes that survive today are the _____ and

    the _____ .

34. A shark's teeth are actually modified _____ .

35. _____ _____ carry oxygen-rich blood from

    an amphibian's _____ to its heart.

36. About 95 percent of all living fish species are _____ .

37. _____ grow throughout a fish's life and can be used to

    estimate the _____ of the fish.

38. All fishes have a(n) _____ _____ that surrounds the spinal cord.

39. In amphibians, one circulatory loop travels to the _____ , and

    another circulatory loop travels to the _____ .

40. The large size of the _____ _____ in the yellow perch's brain indicates the importance of vision to this fish.

41. Teleosts are the most advanced of the _____

    _____ fishes.

42. Individual _____ in the kidneys regulate salt and water balance in an animal's body.

43. Besides breathing with their lungs, frogs, salamanders, and caecilians also

    engage in _____ respiration.

44. Salamanders are members of the order _____ .

45. Urine, undigested food, egg cells, and sperm cells all exit the body of the

    leopard frog through the _____ opening.

46. What is countercurrent flow, and why is it important to a fish?

_____

_____

_____

_____

47. Describe the role of the tympanic membrane in a frog's sense of balance.

_____

_____

_____

_____

48. How do lampreys and hagfishes feed?

_____

_____

_____

49. What information does a bony fish get from its lateral line system?

_____

_____

_____

50. Which group of amphibians does not have the important adaptation of terrestrial vertebrates—legs? Explain.

_____

_____

_____

51. What enables a leopard frog to keep its skin moist? Why is this important?

_____

_____

_____

CHAPTER
34 **VOCABULARY**

Fishes and Amphibians

Complete the crossword puzzle using the clues provided.

ACROSS

2. _____ flow insures that oxygen diffuses into the blood over the entire length of the capillaries.

4. covers the gills on each side of the head

7. The _____ line extends along each side of a bony fish's body and allows the fish to perceive its position and rate of movement.

8. an internal, baglike respiratory organ

11. A swim _____ enables bony fishes to regulate their buoyancy.

DOWN

1. _____ veins carry oxygen-rich blood from the amphibian's lungs to its heart.

3. individual unit of a kidney

5. a fish with a completely symmetrical tail, highly mobile fins, and very thin scales

6. Each gill is made up of rows of fingerlike projections called gill _____ .

9. A _____ slit is an opening at the rear of the cheek cavity.

10. separates the atrium into right and left halves

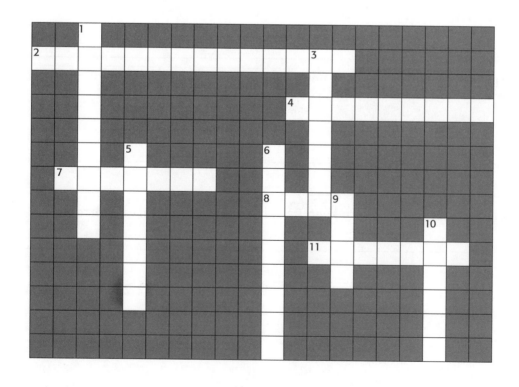

Reptiles and Birds

In the space provided, write the letter of the term or phrase that best completes each statement or best answers each question.

_____ 1. Tuataras are members of the reptile order
- **a.** Chelonia.
- **b.** Squamata.
- **c.** Rhynchocephalia.
- **d.** Crocodilia.

_____ 2. In the raising of their young, crocodiles most closely resemble which of the following vertebrates?
- **a.** turtles
- **b.** lizards
- **c.** snakes
- **d.** birds

_____ 3. All reptiles EXCEPT crocodilians have
- **a.** a partially divided ventricle.
- **b.** alveoli.
- **c.** overlapping scales.
- **d.** relatively small brains.

_____ 4. Which of the following is NOT true of a turtle's shell?
- **a.** Vertebrae are fused to the inside of the carapace.
- **b.** The carapace provides support for muscle attachment.
- **c.** The carapace is always dome shaped.
- **d.** The shell is made of fused plates of bone.

_____ 5. During the Cretaceous period, snakes probably evolved from
- **a.** turtles.
- **b.** lizards.
- **c.** dinosaurs.
- **d.** alligators.

_____ 6. The heat-sensing organ used by timber rattlesnakes to locate a motionless animal in total darkness is called a
- **a.** venom gland.
- **b.** fang.
- **c.** pit organ.
- **d.** rattle.

_____ 7. Which of the following characteristics distinguishes crocodilians from other reptiles?
- **a.** Crocodilians are ectothermic.
- **b.** Crocodilians care for their young after hatching.
- **c.** Crocodilians have dry, watertight skin.
- **d.** Crocodilians have amniotic eggs.

_____ 8. Respiratory efficiency in birds is improved by the presence of
- **a.** air sacs.
- **b.** one-way air flow through the lungs.
- **c.** a completely divided ventricle.
- **d.** All of the above

_____ 9. A bird foot with two forward-facing toes and two backward-facing toes is well adapted for
- **a.** seizing prey.
- **b.** perching, climbing, and holding food.
- **c.** swimming.
- **d.** wading in water.

_____ 10. The problem of reptile sperm and eggs drying out on land is solved by
   a. internal fertilization.
   b. an amniotic egg.
   c. overlapping scales.
   d. Both (a) and (b)

_____ 11. Compared with amphibians, the legs of reptiles are positioned
   a. closer to the head.
   b. farther from the head.
   c. farther apart.
   d. more vertically under the body.

_____ 12. Unlike most reptiles, members of the order Rhynchocephalia are
   a. scaleless.
   b. aquatic.
   c. most active at low temperatures.
   d. endothermic.

_____ 13. The feathers of birds are important for
   a. providing lift for flight.
   b. conserving heat.
   c. insulation.
   d. All of the above

_____ 14. Which of the following is NOT an adaptation for flight?
   a. oil gland
   b. keel for muscle attachment
   c. thin and hollow bones
   d. fused collarbones

_____ 15. The second chamber of the stomach of a bald eagle is known as the
   a. crop.
   b. gizzard.
   c. esophagus.
   d. cloaca.

• • • • • • • • • • • • • • • • •

In the space provided, write the letter of the description that best matches the term or phrase.

_____ 16. ovoviviparous

_____ 17. contour feathers

_____ 18. amniotic egg

_____ 19. Chelonia

_____ 20. alveoli

_____ 21. plastron

_____ 22. Jacobson's organs

_____ 23. snake

_____ 24. grooved teeth

_____ 25. carapace

_____ 26. fangs

a. order that includes turtles and tortoises

b. hollow teeth used to inject venom

c. detect odor of chemicals to help snakes follow prey

d. fertilized eggs develop within a female's body until the eggs hatch

e. direct venom into the victim through bite wounds

f. give an adult bird its shape

g. the bottom part of a turtle or tortoise shell

h. has a jaw that is only loosely connected to its skull

i. the top part of a turtle or tortoise shell

j. grape-shaped chambers that increase the surface area of a reptile's lungs

k. contains both water and food for a developing embryo

Complete each statement by writing the correct term or phrase in the space provided.

27. Crocodiles are the only living reptiles that have a completely divided

    _____ .

28. Because the _____ in a bird's heart is completely divided,

    oxygen-rich and oxygen-poor blood are kept completely _____ .

29. A timber rattlesnake's venom contains _____ that destroy red
    blood cells and cause internal hemorrhaging.

30. The success of lizards and snakes as predators is partially because of how the

    _____ _____ is connected to the skull.

31. Like reptiles, birds lay _____ _____ and have

    _____ on their legs and feet.

Questions 32 and 33 refer to the figure below, which shows the reptilian circulatory system.

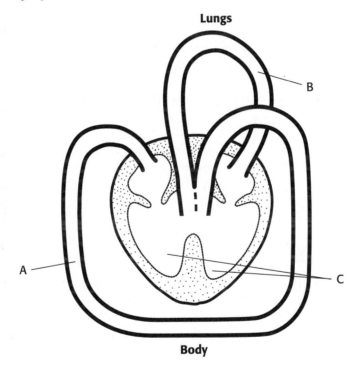

32. Oxygen-poor blood from point *A* enters the _____

    _____ , while oxygen-rich blood from point *B* enters the

    _____ _____ .

33. Arteries carry oxygen-rich blood from the _____ , labeled *C*, to

    the body and oxygen-poor blood to the _____ .

34. Unlike reptiles, birds have _____ and _____ modified into wings.

35. A long, flattened, rounded bill, as found in _____ , is adapted

for _____ .

36. Feathers are modified reptilian _____ .

37. Most reptiles cannot live in very cold regions because they are

_____ .

38. Reptiles, birds, and three species of mammals reproduce by means of a(n)

_____ _____ with a(n) _____ .

This is evidence that they share a(n) _____ _____ .

• • • • • • • • • • • • • • •

Read each question, and write your answer in the space provided.

39. List the four orders of present-day reptiles, and give an example of each.

_____

_____

_____

_____

40. Describe the structure and function of a turtle's shell.

_____

_____

_____

_____

_____

41. Which is more efficient—a bird lung or a reptile lung? Explain.

_____

_____

_____

_____

_____

35 VOCABULARY

Reptiles and Birds

In the space provided, write the letter of the description that best matches the term or phrase.

_____ **1.** amniotic egg

_____ **2.** alveoli

_____ **3.** oviparous

_____ **4.** ovoviviparous

_____ **5.** carapace

_____ **6.** plastron

a. top part of the shell of a turtle

b. chambers located on the inner surface of the lungs

c. encloses the embryo in a watery environment

d. bottom part of the shell of a turtle

e. organisms in which the females retain the eggs within their body until shortly before hatching

f. organisms that produce eggs that hatch outside the mother's body

Complete each statement by writing the correct term or phrase in the space provided.

7. _____ _____ cover the body of young birds and are found beneath the contour feathers of adults.

8. _____ _____ cover the bird's body and give adult birds their shape.

9. A bird protects and waterproofs its feathers by pulling them through its

beak and covering them with oil from its _____

_____ .

CHAPTER

36 **TEST PREP PRETEST**

Mammals

In the space provided, write the letter of the term or phrase that
best completes each statement or best answers each question.

_____ 1. Which of the following is NOT a characteristic of mammals?
  a. hair
  b. specialized teeth
  c. ectothermic metabolism
  d. mammary glands

_____ 2. Monotremes and marsupials have limited geographical distributions
  because of
  a. environmental conditions.
  b. continental drift.
  c. a slow reproductive rate.
  d. competition from placental mammals.

_____ 3. Grizzly bears are able to eat vegetation because they have
  a. a high metabolic rate.
  b. a layer of fat.
  c. a multichambered stomach.
  d. rounded molar teeth with a wrinkled surface.

_____ 4. The whiskers of cats and dogs are used for
  a. insulation.
  b. camouflage.
  c. sensory functions.
  d. a signal to other animals.

_____ 5. All female mammals have
  a. a marsupium.
  b. mammary glands.
  c. pouches.
  d. nipples.

_____ 6. Incisors are used for
  a. biting and cutting.
  b. stabbing and holding.
  c. crushing and grinding.
  d. Both (a) and (b)

_____ 7. Active mammals have increased lung efficiency because of
  a. a divided ventricle.
  b. a high metabolic rate.
  c. smaller and more numerous alveoli.
  d. larger and more numerous alveoli.

Questions 8 and 9 refer to the figure below.

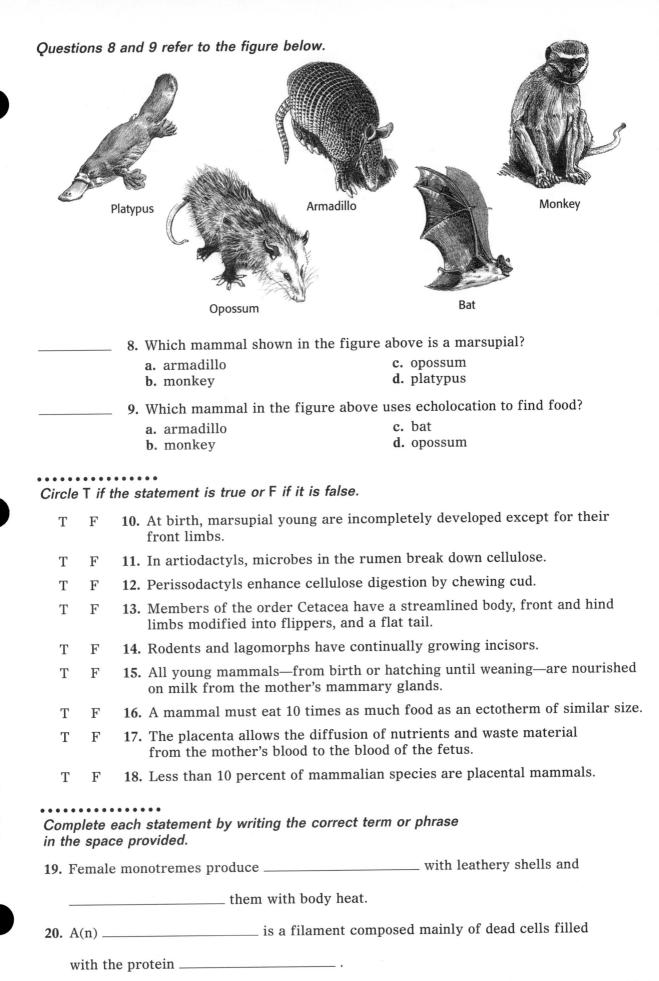

Platypus

Armadillo

Monkey

Opossum

Bat

_____ 8. Which mammal shown in the figure above is a marsupial?
- **a.** armadillo
- **b.** monkey
- **c.** opossum
- **d.** platypus

_____ 9. Which mammal in the figure above uses echolocation to find food?
- **a.** armadillo
- **b.** monkey
- **c.** bat
- **d.** opossum

• • • • • • • • • • • • • • • •

Circle T *if the statement is true or* F *if it is false.*

T F **10.** At birth, marsupial young are incompletely developed except for their front limbs.

T F **11.** In artiodactyls, microbes in the rumen break down cellulose.

T F **12.** Perissodactyls enhance cellulose digestion by chewing cud.

T F **13.** Members of the order Cetacea have a streamlined body, front and hind limbs modified into flippers, and a flat tail.

T F **14.** Rodents and lagomorphs have continually growing incisors.

T F **15.** All young mammals—from birth or hatching until weaning—are nourished on milk from the mother's mammary glands.

T F **16.** A mammal must eat 10 times as much food as an ectotherm of similar size.

T F **17.** The placenta allows the diffusion of nutrients and waste material from the mother's blood to the blood of the fetus.

T F **18.** Less than 10 percent of mammalian species are placental mammals.

• • • • • • • • • • • • • • •

Complete each statement by writing the correct term or phrase in the space provided.

19. Female monotremes produce _____ with leathery shells and

_____ them with body heat.

20. A(n) _____ is a filament composed mainly of dead cells filled

with the protein _____ .

21. The four types of mammalian teeth are _____ ,

_____ , _____ , and _____ .

22. No living animals other than _____ have hair.

23. _____ glands on the chest or abdomen of female mammals produce milk for their young.

24. Grizzly bears rely primarily on their sense of _____ to find food.

25. Respiration in mammals is aided by the _____ , a sheet of muscle at the bottom of the rib cage.

26. The length of time between fertilization and birth is the _____

_____ .

27. The duckbill platypus and two species of echidnas are the only living

_____ .

28. The majority of mammalian species in Australia and New Guinea are

_____ .

29. Placental mammals remain in the _____ through a relatively long gestation period, during which they receive nourishment through the

_____ .

· · · · · · · · · · · · · · · ·

In the space provided, write the letter of the description that best matches the term or phrase.

_____ 30. Order Insectivora

_____ 31. Order Chiroptera

_____ 32. Order Pinnipedia

_____ 33. Order Proboscidea

_____ 34. Order Hyracoidea

_____ 35. Order Endentata

_____ 36. Order Perissodactyla

_____ 37. Order Carnivora

a. includes marine carnivores

b. have elongated nose; largest land animals alive today

c. toothless or have poorly developed teeth without enamel; includes armadillos

d. odd number of toes within their hooves; no rumen

e. only mammals capable of true flight

f. most similar to ancestors of placental mammals; extremely high metabolic rate

g. rabbitlike body; four hooved toes on the front feet and three on the back feet

h. extremely intelligent; excellent sense of smell, vision, and hearing

38. Compare the degree of development and feeding habits of newborn monotremes, marsupials, and placental mammals.

_____

_____

_____

_____

_____

_____

_____

39. Distinguish between the grizzly bear's metabolism and nutrition in summer and winter.

_____

_____

_____

_____

40. Why are there many marsupials in the Australian region and only one in North America?

_____

_____

_____

_____

_____

41. List at least three functions of hair.

_____

_____

42. What ingredients in milk support the rapid growth of young mammals?

_____

_____

36 VOCABULARY

Mammals

*Complete each statement by writing the correct term or phrase
in the space provided.*

1. A(n) _____ is a filament composed mainly of dead cells filled with
   the protein keratin.

2. _____ _____ produce a nutrient-rich energy source
   for nourishing young after their birth.

3. The time when a mother stops nursing her young is called _____ .

4. An organ called the _____ allows the diffusion of nutrients and
   oxygen from the mother's blood into the blood of a developing fetus.

5. The period of time between fertilization and birth is called the _____

   _____ .

6. Mammals with hoofs are classified as _____ .

7. Some mammals regurgitate partly digested food, called _____ ,
   rechew it, and swallow it again for further digestion.

Animal Behavior

Circle T *if the statement is true or* F *if it is false.*

T F **1.** Animals respond only to environmental stimuli.

T F **2.** An animal that does not move as it plays dead is engaging in a defensive behavior.

T F **3.** Natural selection favors traits that contribute to the survival of species.

T F **4.** Nest building in birds is a learned behavior.

T F **5.** Complex behavior has both genetic and learned components.

• • • • • • • • • • • • • • • •

In the space provided, write the letter of the term or phrase that best completes each statement or best answers each question.

_____ **6.** Animal signals are used to

    a. influence an animal's behavior.
    b. solicit play.
    c. attract a mate.
    d. All of the above

_____ **7.** Scientists who question the reasons a behavior exists are asking

    a. a "how" question.
    b. a "why" question.
    c. about the evolution of behavior.
    d. Both (b) and (c)

_____ **8.** Sexual selection is a(n)

    a. innate behavior.
    b. evolutionary mechanism.
    c. behavioral signal.
    d. genetic trait.

_____ **9.** In some animals, extreme traits for acquiring a mate include

    a. horns, antlers, and manes.
    b. the ability to learn.
    c. complex brain structure.
    d. All of the above

_____ **10.** When Konrad Lorenz raised a group of newly hatched goslings, the goslings showed

    a. operant conditioning.   **c.** imprinting.
    b. reasoning.   **d.** classic conditioning.

_____ **11.** Which of the following is NOT a signal?

    a. feeding   **c.** color
    b. sound   **d.** scent

Question 12 refers to the figure at right.

_____ 12. The bird providing food to its young is engaging in
   a. foraging behavior.
   b. parental care.
   c. imprinting.
   d. territorial behavior.

_____ 13. Vocal communication is most developed in
   a. birds.
   b. carnivores.
   c. marine animals.
   d. primates.

_____ 14. Fixed action pattern behaviors are always
   a. innate.
   b. learned.
   c. developed through conditioning.
   d. taught.

_____ 15. Reasoning involves
   a. using past experiences.
   b. analyzing a problem.
   c. developing an insight.
   d. All of the above

• • • • • • • • • • • • • • • • • •

In the space provided, write the letter of the description that best matches the term or phrase.

_____ 16. imprinting

_____ 17. foraging behavior

_____ 18. innate behavior

_____ 19. conditioning

_____ 20. territorial behavior

_____ 21. signal

_____ 22. operant conditioning

_____ 23. habituation

a. locating, obtaining, and consuming food
b. protecting a resource for exclusive use
c. occurs only during a specific period early in an animal's life
d. used to influence another animal's behavior
e. a frequent, harmless stimulus is ignored
f. genetically programmed behavior
g. trial-and-error learning under highly controlled conditions
h. learning by association

• • • • • • • • • • • • • • • •

Complete each statement by writing the correct term or phrase in the space provided.

24. A(n) _____ is an action or series of actions performed by an animal in response to a stimulus.

25. To understand the factors that trigger or control a behavior, a scientist asks a(n)

_____ question.

26. Large antlers in deer, manes in lions, and tusks in walrus are examples of

_____ that attract _____ .

27. Innate behaviors, also called _____ , are genetically programmed.

28. When new male lions in a pride kill cubs of other males, they are demonstrating

a behavior influenced by _____ _____ ,

which favors traits that benefit the _____ and not the species.

29. When rats locked in a box learned to depress a lever to get food, they

demonstrated _____ _____ in a famous study conducted by B. F. Skinner.

30. Birds flying south for the winter are demonstrating _____

_____ .

31. Unique _____ _____ ensure that individuals

do not mate with individuals of another _____ .

32. When Ivan Pavlov's dogs learned to _____ a ringing bell with meat powder, which caused them to salivate, they demonstrated

_____ _____ .

33. Sexual selection as a(n) _____ mechanism was first proposed by Charles Darwin.

• • • • • • • • • • • • • • •
Read each question, and write your answer in the space provided.

34. Explain how imprinting in ducks and geese is influenced by both heredity and learning.

_____

_____

_____

_____

35. Give at least three examples of courtship signals that can be unique to a particular species.

_____

_____

_____

_____

36. What is the difference between a "how" and a "why" behavioral question? Give an example of each.

_____

_____

_____

_____

37. List five types of signals and five methods animals can use to send and receive signals.

_____

_____

_____

_____

_____

38. How did scientists confirm that nest building is innate in Fischer's lovebirds and peach-faced lovebirds?

_____

_____

_____

_____

39. Explain the difference between habituation and classical conditioning.

_____

_____

_____

_____

Animal Behavior

*Complete each statement by writing the correct term or phrase
in the space provided.*

1. A(n) _____ is an action or series of actions performed by an
   animal in response to a stimulus.

2. Genetically programmed behavior (behavior influenced by genes) is often called

   _____ _____ .

3. Innate behavior is called _____ - _____

   _____ _____ when the action always occurs

   in the same way.

4. The development of behaviors through experience is called _____ .

5. Learning by association is called _____ .

6. The ability to analyze a problem and think of a possible solution is called

   _____ .

7. Learning that can occur only during a specific period early in the life of an

   animal and cannot be changed once learned is called _____ .

8. _____ _____ is an evolutionary mechanism in which
   traits that increase the ability of individuals to attract or acquire mates appear with
   increased frequency.

CHAPTER
38 **TEST PREP PRETEST**

Introduction to Body Structure

In the space provided, write the letter of the term or phrase that best completes each statement or best answers each question.

_____ 1. From simplest to most complex, the four levels of structural organization of the human body are as follows:

  a. tissues, cells, organs, organ systems
  b. cells, tissues, organs, organ systems
  c. organ systems, tissues, cells, organs
  d. cells, organs, tissues, organ systems

_____ 2. Types of connective tissue include all of the following EXCEPT

  a. blood.
  b. bone.
  c. fat.
  d. muscle.

_____ 3. Each organ belongs to at least one

  a. tissue.
  b. organ system.
  c. muscle group.
  d. body cavity.

_____ 4. Which of the following would likely occur if endothermy were disrupted?

  a. Body temperature could not be maintained.
  b. Strenuous physical activity would be difficult.
  c. Enzymes would be inactivated.
  d. all of the above

_____ 5. The appendicular skeleton includes bones of the

  a. cranium.
  b. spine.
  c. pelvis.
  d. ribs.

_____ 6. In early development, bone tissue is made mostly of

  a. groups of osteocytes.
  b. bone marrow.
  c. periosteum.
  d. cartilage.

_____ 7. The condition in which bone density is lost and bones become brittle is called

  a. ossification.
  b. osteoporosis.
  c. rheumatoid arthritis.
  d. tendinitis.

_____ 8. Blood cell production begins in the

  a. bone marrow.
  b. Haversian canals.
  c. periosteum.
  d. osteocytes.

_____ **9.** An example of a ball-and-socket joint is the
   a. wrist.
   b. shoulder.
   c. knee.
   d. vertebral column.

_____ **10.** When a muscle contracts, myosin and actin
   a. overlap, shortening each sarcomere.
   b. do not overlap, shortening each sarcomere.
   c. overlap, lengthening each sarcomere.
   d. None of the above

Questions 11 and 12 refer to the figure below, which shows the structure of skin.

_____ **11.** The epidermis is the structure labeled
   a. *A.*   **c.** *C.*
   b. *B.*   **d.** *D.*

_____ **12.** The functional layer of skin is labeled
   a. *A.*   **c.** *C.*
   b. *B.*   **d.** *D.*

• • • • • • • • • • • • • • •

In the space provided, write the letter of the description that best matches the term or phrase.

_____ **13.** compact bone

_____ **14.** spongy bone

_____ **15.** red bone marrow

_____ **16.** yellow bone marrow

_____ **17.** periosteum

_____ **18.** cartilage

_____ **19.** Haversian canals

_____ **20.** osteocytes

a. maintain mineral content of bone

b. template tissue for bone formation

c. site of blood cell production

d. channels containing blood vessels in concentric rings of compact bone

e. dense bone that provides a great deal of support

f. tough membrane surrounding bones

g. site of fat storage

h. loosely structured bone

Copyright © by Holt, Rinehart and Winston. All rights reserved.

Complete each statement by writing the correct term or phrase in the space provided.

21. Bones of the skull, spine, and rib cage make up the _____

_____ .

22. The excretory system includes the kidneys, _____

_____ , ureters, and the _____ .

23. The body maintains a constant temperature because of _____ , which enables homeostasis.

24. Osteoporosis can be delayed or prevented by exercise and a diet containing

ample _____ .

25. A muscle fiber contains many bundles of cylindrical structures called

_____ .

26. In myofibrils, actin filaments are anchored at _____

_____ .

27. When oxygen is plentiful, the ATP used to power muscle contractions is supplied

by _____ processes. As oxygen becomes depleted, ATP is

supplied by _____ processes.

28. Muscle fatigue and soreness result when ATP _____ exceeds

ATP _____ .

29. _____ , a pigment found in the epidermis, absorbs UV radiation and helps determine skin color.

30. Hair consists mostly of dead cells filled with the protein called

_____ , the same protein that makes the skin tough and waterproof.

31. _____ _____ in the dermis help regulate body temperature, either by radiating heat into the air or by helping to insulate the body.

32. The most common type of skin cancer originates in cells of the

_____ that do not produce pigments.

33. Skin cancers that occur in pigment-producing skin cells are called

_____ _____ .

34. Acne is caused by excessive secretion of _____ , an oily secretion that lubricates the skin, by oil glands.

Read each question, and write your answer in the space provided.

35. Using a cardiac muscle cell as an example, describe how the cell is the foundation of tissues, organs, and organ systems.

_____

_____

36. Relate the process of bone formation to the importance of eating a mineral-rich diet.

_____

_____

_____

37. Describe the three main types of joints, and give an example of each.

_____

_____

_____

_____

38. Explain how rheumatoid arthritis can affect freely movable joints.

_____

_____

_____

39. Describe the interaction of myosin and actin during a muscle contraction.

_____

_____

_____

40. Differentiate between the epidermis, the dermis, and subcutaneous tissue.

_____

_____

_____

41. What is melanin, and where is it produced?

_____

_____

Name_____ Date _____ Class _____

Introduction to Body Structure

In the space provided, write the letter of the description that best matches the term or phrase.

_____ 1. epithelial tissue

_____ 2. nervous tissue

_____ 3. muscle tissue

_____ 4. connective tissue

_____ 5. body cavities

a. carries information throughout the body

b. provides support, protection, and insulation

c. fluid-filled spaces that contain major body organs

d. enables the movement of body structures

e. lines most body surfaces

Complete each statement by writing the correct term or phrase in the space provided.

6. The _____ _____ includes bones of the skull, spine, ribs, and sternum.

7. The _____ _____ includes bones of the arms, legs, pelvis, and shoulder.

8. _____ _____ is soft tissue inside bones that begins the manufacture of blood cells.

9. Bones are surrounded by a tough exterior membrane called the _____ that contains blood vessels that supply nutrients to bones.

10. _____ _____ are hollow channels that contain blood vessels that enter bone through the periosteum.

11. Bone cells called _____ maintain the mineral content of bone.

12. _____ , meaning "porous bone," is a disease that results in brittle bones and that affects more women than men.

13. A junction between two or more bones is called a(n) _____ .

14. Bones of a joint are held together by strong bands of connective tissue called

_____ .

Use the terms from the list below to fill in the blanks in the following passage.

actin   myofibrils   sarcomere

extensor   myosin   tendons

flexor

Most skeletal muscles are attached to bones by strips of dense connective

tissue called (15) _____ . One muscle in a pair of muscles

pulls a bone in one direction, and the other muscle pulls the bone in the

opposite direction. A(n) (16) _____ muscle causes a joint to

bend, and a(n) (17) _____ muscle causes a joint to straighten.

Muscle tissue contains large amounts of protein filaments called

(18) _____ and (19) _____ , which enable

muscles to contract. Each muscle fiber is made of small cylindrical structures

called (20) _____ , which have alternating light and dark

bands when viewed under a microscope. In the center of each light band is

a Z line, which anchors actin filaments. The area between the two Z lines is

called a(n) (21) _____ , the functional unit of muscle

contraction.

In the space provided, write the letter of the description that best matches the term or phrase.

_____ 22. epidermis

_____ 23. keratin

_____ 24. melanin

_____ 25. dermis

_____ 26. subcutaneous tissue

_____ 27. hair follicle

_____ 28. sebum

a. protein that makes skin tough and waterproof

b. functional layer of skin that lies just beneath the epidermis

c. the outermost layer of skin

d. oil secretion that lubricates the skin

e. made mostly of fat

f. pigment that helps determine skin color

g. produces individual hairs

Name_____ Date _____ Class _____

Introduction to Body Structure

Use the figure below to answer questions 1–4 below.

A   B

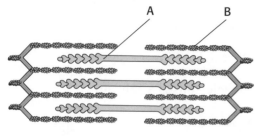

Step 1

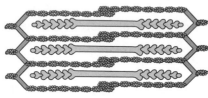

Step 2

Step 3

Read each question, and write your answer in the space provided.

1. The striped pattern of a sarcomere is formed by two different protein filaments, labeled *A* and *B*. Identify and briefly describe these two filaments.

_____

_____

_____

2. How do these two protein filaments interact during muscle contraction?

_____

_____

3. Explain what is occurring in Step 2.

_____

_____

_____

_____

4. What substance is needed to go from Step 3 back to Step 1? Explain.

_____

_____

In the space provided, write the letter of the each labeled bone in the figure at right that matches the name of each bone.

_____ 5. fibula

_____ 6. ulna

_____ 7. pelvic girdle

_____ 8. scapula

_____ 9. sternum

_____ 10. spine

_____ 11. femur

_____ 12. carpals

_____ 13. metatarsals

_____ 14. patella

_____ 15. humerus

_____ 16. skull

_____ 17. metacarpals

_____ 18. tibia

_____ 19. radius

_____ 20. clavicle

_____ 21. tarsals

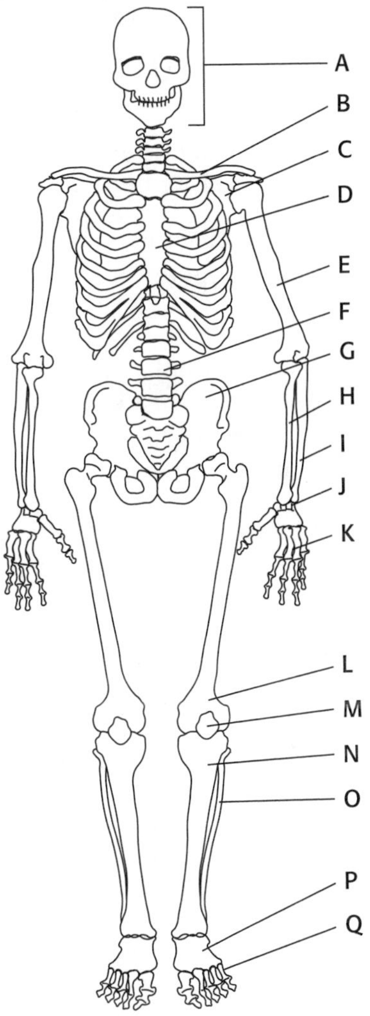

Read each question, and write your answer in the space provided.

22. What are the primary functions of the skeleton?

_____

_____

_____

_____

23. Distinguish between the axial skeleton and the appendicular skeleton.

_____

_____

_____

_____

Circulatory and Respiratory Systems

In the space provided, write the letter of the term or phrase that
best completes each statement or best answers each question.

_____ 1. The circulatory system transports
    a. oxygen.
    b. nutrients.
    c. hormones.
    d. All of the above

_____ 2. The actual exchange of materials between the blood and the cells of the
    body occurs in the
    a. arteries.
    b. arterioles.
    c. veins.
    d. capillaries.

_____ 3. When fluids leak out of the cardiovascular system, they are returned by the
    a. respiratory system.
    b. lymphatic system.
    c. endocrine system.
    d. digestive system.

_____ 4. Blood type is determined by the presence or absence of
    a. A and B antigens dissolved in the blood plasma.
    b. A and O antigens on the surface of red blood cells.
    c. A and B antigens on the surface of red blood cells.
    d. A and B antigens on the surface of white blood cells.

_____ 5. The blood pumped from the heart to the lungs is transported through
    a. the pulmonary circulation loop.
    b. the systemic circulation loop.
    c. both the pulmonary and systemic loops.
    d. neither the pulmonary nor the systemic loop.

_____ 6. As the left ventricle contracts, the blood is prevented from moving back
    into the left atrium by
    a. a one-way valve.
    b. the superior vena cava.
    c. the inferior vena cava.
    d. the septum.

_____ 7. The natural pacemaker of the heart is the
    a. aorta.
    b. sinoatrial node.
    c. coronary artery.
    d. superior vena cava.

_____ 8. Blood is pumped from the left ventricle to the
    a. lungs.
    b. body tissues.
    c. left atrium.
    d. right ventricle.

_____ 9. A recording of the electrical changes that occur in the heart each time it
    contracts is called a(n)
    a. pulse reading.
    b. sphygmomanometer.
    c. wave contraction.
    d. electrocardiogram.

_____ 10. When part of the heart muscle dies from lack of oxygen and the entire organ stops working, the result is
   a. anemia.
   b. a heart attack.
   c. a stroke.
   d. a seizure.

_____ 11. The structure that prevents food and liquid from entering the trachea is called the
   a. pharynx.
   b. larynx.
   c. alveolus.
   d. epiglottis.

_____ 12. During inhalation,
   a. the diaphragm contracts and moves downward, and the rib cage moves upward and outward.
   b. the diaphragm expands and moves upward, and the rib cage moves upward and outward.
   c. the diaphragm contracts and moves downward, and the rib cage moves downward and inward.
   d. the diaphragm and the rib cage return to their normal resting positions.

_____ 13. Hemoglobin contains four atoms of iron that bind reversibly with
   a. carbonic acid.
   b. oxygen.
   c. bicarbonate ions.
   d. water.

_____ 14. The respiratory disease in which the bronchioles of the lungs become constricted because of their sensitivity to certain stimuli in the air is called
   a. lung cancer.
   b. emphysema.
   c. asthma.
   d. tuberculosis.

· · · · · · · · · · · · · · · ·

In the space provided, write the letter of the description that best matches the term or phrase.

_____ 15. erythrocytes

_____ 16. artery

_____ 17. white blood cells

_____ 18. vein

_____ 19. plasma

_____ 20. anemia

_____ 21. type O blood

_____ 22. type AB blood

_____ 23. platelets

_____ 24. fibrin

a. the largest blood cells; also known as leukocytes

b. the portion of blood containing metabolites, wastes, salts, proteins, and water

c. red blood cells

d. a condition characterized by reduced oxygen-carrying capacity of the blood

e. cell fragments needed to form blood clots

f. universal recipient

g. blood vessel that carries blood away from the heart

h. universal donor

i. blood vessel that returns blood to the heart

j. the sticky protein threads that function in blood clotting

Complete each statement by writing the correct term or phrase in the space provided.

25. _____ _____ , such as carbon dioxide, are transported by the circulatory system to excretory organs and tissues.

26. The large veins called _____ _____ carry

    oxygen-poor blood from the body into the _____

    _____ of the heart.

27. White blood cells, or _____ , are the primary cells of the immune system.

28. The _____ _____ loop transports blood from the left side of the heart to body tissues and then to the right side of the heart.

29. The heart receives blood in the two _____ and pumps blood

    away using the two _____ .

30. The _____ _____ in the right atrium initiates each heart contraction.

31. The pressure that is measured during relaxation of the heart is called

    _____ pressure, and the pressure that is measured upon

    contraction of the ventricles is called _____ pressure.

32. When blood pressure becomes too high, a condition called

    _____ results.

33. The _____ are suspended in the chest cavity and are bound on the sides by the ribs and on the bottom by the diaphragm.

34. The alveoli are connected to the bronchi by a network of tiny tubes called

    _____ .

35. During breathing, _____ occurs when the diaphragm and rib cage return to their relaxed position.

36. The involuntary regulation of breathing is due to special _____ in the brain and circulatory system.

37. The presence of carbon dioxide in the blood makes the blood more

    _____ .

38. _____ _____ and _____ are diseases of the respiratory system that have been linked to cigarette smoking.

39. List the main plasma ions, and describe their functions.

_____

_____

_____

40. How does lymph move through the lymphatic system?

_____

_____

_____

41. How are oxygen and carbon dioxide transported in the blood?

_____

_____

_____

_____

42. What is one factor that stimulates receptors in the brain, causing an increase in the breathing rate?

_____

_____

Questions 43 and 44 refer to the table below.

Blood type	Antigen on red blood cells	Can give blood to
A	A	A, AB
B	B	B, AB
AB	A and B	AB
O	none	A, B, AB, O

43. Explain why a person with type A blood cannot receive a blood transfusion from a donor with type B blood.

_____

_____

44. Explain why type O blood can be donated in a blood transfusion regardless of the recipient's blood type.

_____

_____

Circulatory and Respiratory Systems

Use the terms from the list below to fill in the blanks in the following passage.

ABO blood group system	lymphatic system	Rh factor
anemia	plasma	valves
arteries	platelets	veins
capillaries	red blood cells	white blood cells

Blood circulation describes the route blood takes as it leaves and then returns to the heart. (1) _____ are blood vessels that carry blood away from the heart. From the arteries, the blood passes into a network of smaller arteries called arterioles. Eventually, the blood is pushed through to the (2) _____ , which are tiny blood vessels that allow the exchange of gases, nutrients, hormones, and other molecules traveling in the blood. After leaving the capillaries, the blood flows into small vessels called venules before emptying into larger vessels called (3) _____ , which are blood vessels that carry the blood back to the heart. Most veins have one-way (4) _____ , which are flaps of tissue that prevent the backflow of blood.

The (5) _____ _____ is a system of the body that collects and recycles fluids that leak from the circulatory system. It is also involved in fighting infections.

About 60 percent of the total volume of blood is (6) _____ . Most of the cells that make up blood are (7) _____ _____ _____ . An abnormality in the size, shape, color, or number of these cells results in (8) _____ , which means that the oxygen-carrying ability of the blood is reduced.

(9) _____ _____ _____

are cells whose primary job is to defend the body against disease.

(10) _____ play an important role in the clotting of blood.

Occasionally an injury or disorder is so serious that a person must receive

blood or blood components from another person. The blood of the recipient and

that of the donor must be compatible—their blood types must match. Under the

(11) _____ _____ _____

_____ , the primary blood types are A, B, AB, and O. The

letters *A* and *B* refer to proteins on the surface of red blood cells that act as

antigens. Another important antigen is called the (12) _____

_____ .

In the space provided, write the letter of the description that
best matches the term or phrase.

_____ 13. atrium

_____ 14. ventricle

_____ 15. vena cava

_____ 16. aorta

_____ 17. coronary arteries

_____ 18. sinoatrial node

_____ 19. blood pressure

_____ 20. pulse

_____ 21. heart attack

_____ 22. stroke

_____ 23. pharynx

_____ 24. larynx

_____ 25. trachea

_____ 26. bronchi

_____ 27. alveoli

_____ 28. diaphragm

a. the first arteries to branch from the aorta

b. chamber that pumps blood away from the heart

c. a small cluster of cardiac muscle cells; initiate heart contraction

d. the force exerted by blood as it moves through the blood vessels

e. the largest artery in the body

f. a series of pressure waves within an artery

g. chamber that receives blood returning to the heart

h. when an area of the heart muscle stops working

i. vessel that collects oxygen-poor blood from the body

j. when an area of the brain does not receive enough blood

k. a long, straight tube in the chest cavity through which air passes

l. two small tubes that lead to the lungs

m. the voice box

n. a muscle at the bottom of the rib cage

o. a muscular tube in the upper throat

p. air sacs where gases are exchanged

Name_____ Date _____ Class _____

Digestive and Excretory Systems

In the space provided, write the letter of the term or phrase that best completes each statement or best answers each question.

_____ 1. A substance needed by the body for energy, growth, repair, and maintenance is called a(n)
    a. fatty acid.   **c.** nutrient.
    b. amino acid.   **d.** calorie.

_____ 2. The unit used to measure the amount of energy available in food is the
    a. Calorie.   **c.** gram.
    b. meter.   **d.** liter.

_____ 3. More than one-half of one day's calories should come from foods high in
    a. complex carbohydrates.   **c.** protein.
    b. lipids.   **d.** simple carbohydrates.

_____ 4. A diet high in saturated fats can be linked to
    a. diabetes.
    b. anorexia nervosa.
    c. bulimia.
    d. high blood cholesterol levels.

_____ 5. According to the USDA food guide pyramid, a person should obtain the most servings per day from
    a. fruits.
    b. breads, cereals, rice, and pasta.
    c. fats, oils, and sweets.
    d. milk, yogurt, and cheese.

_____ 6. Amylases begin the breakdown of carbohydrates into
    a. fatty acids.   **c.** amino acids.
    b. polypeptides.   **d.** simple sugars.

_____ 7. In the stomach, single protein strands are cut into smaller amino acid chains by the digestive enzyme called
    a. amylase.   **c.** lipase.
    b. pepsin.   **d.** gastrin.

_____ 8. The products of digestion are absorbed into the bloodstream through the
    a. villi and microvilli of the small intestine.
    b. rectum of the large intestine.
    c. gall bladder.
    d. sphincter of the stomach.

_____ 9. Bile, which emulsifies fat globules, is produced by the
    a. pancreas.   **c.** liver.
    b. gallbladder.   **d.** duodenum.

10. During the metabolism of proteins and nucleic acids, the toxic waste product that is formed is

a. urea.
b. urine.
c. ammonia.
d. carbon dioxide.

11. The end result of the filtering, absorption, and secretion processes in the nephrons is

a. water.
b. carbon dioxide.
c. urine.
d. urea.

Questions 12–14 refer to the figure at right.

12. The blood-filtering unit in the figure is called a(n)

a. villus.
b. nephron.
c. urethra.
d. microvillus.

13. The structure labeled *A* is called the

a. loop.
b. glomerulus.
c. renal tubule.
d. Bowman's capsule.

14. The structure labeled *C* is called the

a. loop of Henle.
b. glomerulus.
c. renal tubule.
d. Bowman's capsule.

15. Urine leaves the bladder and exits the body through a tube called the

a. urethra.
b. ureter.
c. kidney.
d. nephron.

. . . .

Circle T *if the statement is true or* F *if it is false.*

T F **16.** A calorie is the amount of heat energy required to raise the temperature of 1 kg of water 1°C.

T F **17.** A teen's body is able to make all of the amino acids necessary to make proteins.

T F **18.** Vitamins and minerals are required in our diets even though they do not provide energy.

T F **19.** Saliva is secreted into the mouth by three pairs of salivary glands.

T F **20.** Once food passes through the pharynx, it enters the esophagus.

T F **21.** During emulsification, bile salts break down large fat globules into smaller fat droplets.

T F **22.** Gastric juice regulates the synthesis of hydrochloric acid, permitting it to be made only when the pH in the stomach is higher than about 1.5.

T F **23.** Ulcers in the stomach or small intestine are caused by extreme obesity.

T F **24.** The process of breaking down food into molecules that the body can use is called digestion.

T F   **25.** Microvilli on the surface of the villi greatly increase the internal surface area of the large intestine.

T F   **26.** Urea is a principal component of urine.

T F   **27.** The kidney is composed of two blood-filtering units called nephrons.

T F   **28.** The fluid inside the Bowman's capsule is called filtrate.

T F   **29.** A transplanted kidney is never rejected by the recipient's body.

· · · · · · · · · · · · · · ·

Complete each statement by writing the correct term or phrase in the space provided.

30. Excess carbohydrates are stored as _____ in the liver and in some muscle tissue.

31. Trace elements are _____ present in the body in small amounts.

32. Successive rhythmic waves of contraction of the smooth muscles around the

esophagus, called _____ _____ , move the food toward the stomach.

33. Of the four functions of the digestive system, the last to occur is the process of

getting rid of undigested molecules and _____ .

34. Although you can live several weeks without food, you can survive only a few

days without _____ .

35. The wall of the large intestine absorbs mostly _____

_____ and _____ .

36. _____ _____ and some water are excreted by the lungs when you exhale.

37. The Bowman's capsule is connected to a long, narrow tube called the

_____ _____ .

38. The renal tubule empties into a larger tube called a(n) _____

_____ .

39. _____ _____ is a procedure for filtering the blood using a machine.

Read each question, and write your answer in the space provided.

40. Describe the connection between heart disease and the USDA food pyramid's daily serving recommendation for fats.

_____

_____

41. What keeps the acid-soaked food in the stomach from reentering the esophagus?

_____

_____

42. How do the liver and the pancreas differ from other digestive organs?

_____

_____

_____

43. Describe the similarities and differences between a mineral and a vitamin.

_____

_____

_____

44. Name two organs other than the kidney that are involved in excretion, and describe what each organ excretes.

_____

_____

_____

45. Briefly describe the three phases that occur as blood flows through a nephron.

_____

_____

_____

_____

_____

46. Relate the role of water in maintaining a healthy body.

_____

_____

_____

Digestive and Excretory Systems

*Complete each statement by writing the correct term or phrase
in the space provided.*

1. A(n) _____ is a substance the body needs for energy, growth, repair, and maintenance.

2. The process of breaking down food into molecules the body can use is called

   _____ .

3. A(n) _____ is the amount of heat energy required to raise the temperature of 1 g of water 1°C (1.8°F).

4. Organic substances that are necessary, in trace amounts, for the normal metabolic

   functioning of the body are called _____ .

5. Inorganic substances that are necessary to make certain body structures and substances, to continue normal nerve and muscle function, and to maintain osmotic

   balance are called _____ .

*In the space provided, write the letter of the description that
best matches the term or phrase.*

_____ 6. amylase

_____ 7. esophagus

_____ 8. peristaltic contraction

_____ 9. pepsin

_____ 10. ulcer

_____ 11. lipase

_____ 12. villi

_____ 13. colon

_____ 14. hepatitis

a. successive rhythmic waves of smooth muscle contraction

b. the large intestine

c. fine, fingerlike projections in the small intestine

d. pancreatic enzyme that digests fat

e. a long tube that connects the mouth to the stomach

f. a digestive enzyme secreted by the stomach

g. inflammation of the liver

h. a hole in the wall of the stomach or small intestine

i. an enzyme in saliva that breaks down carbohydrates

Use the terms from the list below to fill in the blanks in the following passage.

excretion   ureters   urinary bladder
nephrons   urethra   urine
urea

(15) _____ is the process that rids the body of toxic

chemicals, excess water, salts, and carbon dioxide and that maintains osmotic

and pH balances. The organs of excretion are the lungs, the kidneys, and the

skin. The liver plays a role in excretion because it converts ammonia, a toxic

nitrogen-containing waste, to (16) _____ , which is a much

less toxic waste.

(17) _____ are tiny tubes in the kidneys with cup-shaped

capsules surrounding a tight ball of capillaries that filter wastes from

the blood. These tubes retain useful molecules, and they produce

(18) _____ . (19) _____ are tubes that

carry urine from the kidney into the (20) _____

_____ . Urine leaves the body through a tube called the

(21) _____ .

The Body's Defenses

*In the space provided, write the letter of the term or phrase that
best completes each statement or best answers each question.*

_____ 1. White blood cells that kill bacteria by engulfing them and then releasing
chemicals that kill both the bacteria and themselves are

    a. macrophages.
    b. neutrophils.
    c. natural killer cells.
    d. helper T cells.

_____ 2. The temperature response is helpful in fighting bacteria because

    a. higher temperatures promote the activation of cellular enzymes.
    b. lower temperatures promote the activation of cellular enzymes.
    c. disease-causing bacteria do not grow well at high temperatures.
    d. disease-causing bacteria do not grow well at low temperatures.

_____ 3. The four main types of cells involved in the immune response are

    a. macrophages, cytotoxic T cells, helper T cells, and B cells.
    b. macrophages, red blood cells, helper T cells, and neutrophils.
    c. macrophages, red blood cells, helper T cells, and natural
       killer cells.
    d. red blood cells, cytotoxic T cells, helper T cells, and B cells.

*Questions 4–7 refer to the figure at right, which
shows an immune response.*

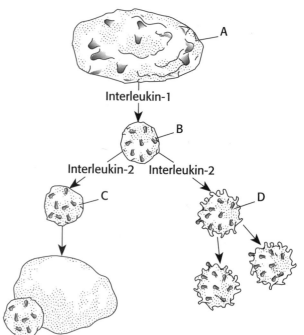

_____ 4. The cell labeled *A* is a

    a. macrophage.
    b. helper T cell.
    c. cytotoxic T cell.
    d. B cell.

_____ 5. The cell labeled *B* is a

    a. plasma cell.
    b. cytotoxic T cell.
    c. helper T cell.
    d. B cell.

_____ 6. The cell labeled *C* is a

    a. cytotoxic T cell.
    b. macrophage.
    c. plasma cell.
    d. memory cell.

_____ 7. The cells produced by the
cell labeled *D*

    a. release antibodies.
    b. kill virus-infected cells.
    c. engulf viruses.
    d. secrete interleukin-2.

8. Smallpox was eradicated by
   a. more-sanitary living conditions.
   b. water purification programs.
   c. vaccination.
   d. None of the above

9. Severe pain and inflammation of the joints is a symptom of the autoimmune disease called
   a. Graves' disease.
   b. AIDS.
   c. type I diabetes.
   d. rheumatoid arthritis.

10. In an autoimmune disease,
   a. a pathogen is immune to antigens.
   b. a pathogen circulates in the blood.
   c. the body attacks its own cells.
   d. the immune system collapses.

11. Which of the following is NOT a way that HIV can be transmitted?
   a. sexual intercourse with an infected person
   b. injecting intravenous drugs with hypodermic needles contaminated with HIV-infected white blood cells
   c. blood transfusions in areas where tests for HIV are unavailable
   d. insect bites

• • • • • • • • • • • • • • •
Circle T *if the statement is true or* F *if it is false.*

T   F   **12.** The skin is a nonspecific defense against invasion by pathogens.

T   F   **13.** Mucous membranes serve as a barrier to pathogens, and they produce chemical defenses.

T   F   **14.** Because of your body's defenses, it is impossible for pathogens to enter your body when you eat or breathe.

T   F   **15.** Red blood cells are called macrophages.

T   F   **16.** Natural killer cells are one of the body's best defenses against cancer.

T   F   **17.** When complement proteins encounter pathogens, they form a membrane attack complex.

T   F   **18.** The inflammatory response slows healing because of the intense swelling that accompanies it.

T   F   **19.** An immune response is activated as soon as a pathogen reaches the skin.

T   F   **20.** Cytotoxic T cells label invaders for later destruction by macrophages.

T   F   **21.** Infected cells display some of the pathogen's antigens on their surface.

T   F   **22.** Vaccination triggers an immune response against the pathogen without symptoms of infection.

T   F   **23.** In an allergic reaction, the body responds to a harmless substance as if the substance were a pathogen.

T   F   **24.** HIV attacks and kills helper T cells.

T   F   **25.** HIV can be transmitted by drinking from a water fountain used by an infected person.

Complete each statement by writing the correct term or phrase in the space provided.

26. Cells lining the bronchi and bronchioles in the respiratory tract secrete a layer

of _____ that traps pathogens before they can enter the lungs.

27. A chemical defense called the _____

_____ consists of proteins that circulate in the blood.

28. Once a virus has entered your body and has begun to infect cells, macrophages

release the protein called _____ .

29. Helper T cells activate _____ T cells and

_____ cells.

30. To combat a viral invader, helper T cells release the protein

_____ .

31. _____ is the introduction of a dead or modified pathogen into the body.

32. _____ _____ is an autoimmune disease in which the insulating material surrounding nerve cells is destroyed.

33. Symptoms of allergic reactions, including swelling, itchy eyes, and nasal

congestion, are caused by the release of _____ .

34. Because of _____ _____ , influenza viruses can reinfect a person even after memory cells have produced immunity.

35. An enzyme that prevents viruses from making proteins and RNA is the result of

a nonspecific defense called _____ .

Read each question, and write your answer in the space provided.

36. List Koch's postulates. How do biologists use Koch's postulates?

_____

_____

_____

_____

_____

37. How can people decrease their exposure to pathogens?

_____

_____

38. What causes the pus that accompanies some infections?

_____

_____

_____

39. Name three kinds of white blood cells involved in nonspecific defenses. Where are they found, and how does each type attack pathogens?

_____

_____

_____

_____

_____

40. How does a person become immune to a pathogen?

_____

_____

_____

_____

_____

41. Name three autoimmune diseases, and describe their symptoms.

_____

_____

_____

42. List five ways that diseases are transmitted to humans.

_____

_____

_____

_____

43. How can one reduce the risk of HIV infection?

_____

_____

_____

_____

CHAPTER
41 VOCABULARY

The Body's Defenses

In the blanks provided, fill in the letters of the term or phrase being described.

1. a disease-causing agent
   _ A_ _ _ _ _ _

2. layers of epithelial tissue that serve as barriers to pathogens and produce chemical defense
   M_ _ _ _ _ M_ _ _ _ _ _ _

3. a series of events that suppress infection
   _ _ _ L_ _ _ _ _ _ _ _ E_ _ _ _

4. chemical that causes local blood vessels to dilate
   _ _ _ T_ _ _ _ _ _

5. a defense mechanism with 20 different proteins
   _ _ _ _ _ M_ _ _ Y_ _ _ _

6. a protein released by cells infected with viruses
   _ _ _ _ _ F_ _ _ _

7. a white blood cell that releases chemicals that kill pathogens
   N_ _ _ _ _ _ _ _

8. a white blood cell that ingests and kills pathogens
   _ _ R_ _ _ _ _ _

9. destroys an infected cell by puncturing its membrane
   N_ _ _ _ _ _ _ _ _ _ L_ _ _ _ _ L

Use the terms from the list below to fill in the blanks in the following passage.

antibodies   B cells   helper T cells
antigens   cytotoxic T cells plasma cells

White blood cells are produced in bone marrow and circulate in blood and

lymph. Four main kinds of white blood cells are involved in the immune

response. Macrophages consume pathogens and infected cells.

(10) _____ _____ _____

attack and kill infected cells. (11) _____

_____ label invaders for later destruction by macrophages.

(12) _____ _____ _____

activate both cytotoxic T cells and B cells.

An infected body cell will display (13) _____ of an invader

on its surface. These are substances that trigger an immune response.

In an immune response, B cells divide and develop into

(14) _____ _____ , which release special

defensive proteins into the blood. These special proteins are called

(15) _____ .

In the space provided, write the letter of the description that best matches the term or phrase.

_____ **16.** Koch's postulates

_____ **17.** immunity

_____ **18.** vaccination

_____ **19.** vaccine

_____ **20.** antigen shifting

_____ **21.** autoimmune disease

_____ **22.** AIDS

_____ **23.** HIV

_____ **24.** allergy

_____ **25.** CD4

a. body's overreaction to a normally harmless antigen

b. when the body launches an immune response against its own cells

c. a medical procedure used to produce resistance

d. the virus that causes AIDS

e. a guide for identifying specific pathogens

f. resistance to a particular disease

g. a solution that contains a dead or modified pathogen that can no longer cause disease

h. acquired immunodeficiency syndrome

i. when a pathogen produces a new antigen that the immune system does not recognize

j. receptor protein recognized by HIV

_____ Date _____ Class _____

Nervous System

In the space provided, write the letter of the term or phrase that best completes each statement or best answers each question.

_____ 1. The difference in electrical charge across a cell membrane is called
    a. a synapse.
    b. conduction.
    c. the membrane potential.
    d. synaptic transmission.

_____ 2. The brain and the spinal cord make up the
    a. central nervous system.
    b. peripheral nervous system.
    c. autonomic nervous system.
    d. None of the above

Questions 3–6 refer to the figure below, which shows the structure of a typical neuron.

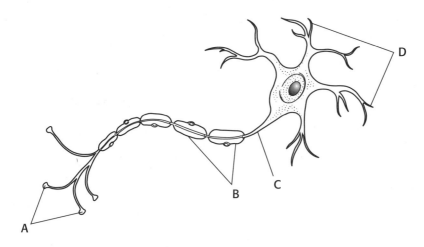

_____ 3. The structures labeled *A*
    a. are axon terminals.
    b. transmit information to other cells.
    c. release neurotransmitters.
    d. All of the above

_____ 4. The structures labeled *B* are
    a. axon terminals.
    b. nodes of Ranvier.
    c. dendrites.
    d. myelin sheaths.

_____ 5. The structure labeled *C* is a(n)
    a. dendrite.
    b. axon.
    c. axon terminal.
    d. cell body.

_____ 6. The structures labeled *D* are
    a. dendrites.
    b. axons.
    c. axon terminals.
    d. nodes of Ranvier.

_____ 7. The peripheral nervous system connects the body to the

    a. upper brain stem.
    b. hypothalamus.
    c. brain and the spinal cord.
    d. autonomic nervous system.

_____ 8. During a knee-jerk reflex, the nerve impulse is received by the

    a. brain.
    b. spinal cord.
    c. spinal cord and then the brain.
    d. thalamus.

_____ 9. In times of stress, the division of the autonomic nervous system that dominates is the

    a. sympathetic division.
    b. parasympathetic division.
    c. motor division.
    d. sensory division.

_____ 10. Light entering the eye stimulates

    a. hair cells in the retina.
    b. the optic nerve.
    c. mechanoreceptors.
    d. rods and cones in the retina.

_____ 11. Taste is detected by

    a. thermoreceptors.
    b. photoreceptors.
    c. chemoreceptors.
    d. olfactory receptors.

_____ 12. When cocaine interferes with reuptake receptors on a presynaptic neuron, the

    a. postsynaptic cell is overstimulated.
    b. number of neurotransmitter receptors decreases.
    c. excess neurotransmitters remain in the synaptic cleft.
    d. All of the above

_____ 13. The body's natural painkillers include

    a. stimulants.
    b. acetylcholine.
    c. enkephalins.
    d. hallucinogens.

_____ 14. Alcohol consumption can

    a. alter neurons throughout the nervous system.
    b. affect normal brain function.
    c. cause abnormalities in the circulatory system.
    d. All of the above

_____ 15. Repeated drug use that alters normal functioning of neurons and synapses results in

    a. addiction.
    b. withdrawal.
    c. tolerance.
    d. None of the above

_____ 16. Myelin sheaths

    a. are found on all neurons.
    b. increase the speed of nerve impulses.
    c. decrease the speed of nerve impulses.
    d. All of the above

Questions 17–20 refer to the figure at right.

_____ **17.** The structures labeled *A*, *B*, and *C* regulate

   a. heart rate and sleep.
   b. body temperature.
   c. breathing rate.
   d. All of the above

_____ **18.** The structure labeled *D* is involved in

   a. balance and posture.
   b. maintaining homeostasis.
   c. spinal reflexes.
   d. learning, memory, and perception.

_____ **19.** The structure labeled *E* is the

   a. thalamus.
   b. corpus callosum.
   c. brain stem.
   d. cerebrum.

_____ **20.** The structure labeled *F* is the

   a. thalamus.
   b. hypothalamus.
   c. cerebellum.
   d. cerebral cortex.

• • • • • • • • • • • • • • • •

Complete each statement by writing the correct term or phrase in the space provided.

21. A bundle of axons is called a(n) _____ .

22. The _____ _____ of a neuron is negative because there are more positively charged ions outside the cell than inside the cell.

23. A(n) _____ _____ is a local reversal of polarity inside a neuron.

24. During synaptic transmission, a presynaptic neuron releases a(n)

_____ into the synaptic _____ .

25. At a synapse, a neurotransmitter may _____ or

_____ a postsynaptic cell.

26. After a nerve impulse has passed, _____ ions flow out of the

axon and the membrane potential becomes _____ again.

27. In the spinal cord, cell bodies of neurons make up the _____

matter, while axons make up the _____ matter.

28. The _____ nervous system contains neurons that connect the brain and the spinal cord to the rest of the body.

29. The coiled inner ear structure that converts sound waves to nerve impulses is

called the _____ .

30. Pain receptors are located on all tissues and organs except the

_____ .

31. Auditory information is processed in the _____

_____ of the brain.

32. _____ _____ alter the functioning of the
central nervous system, and they are often addictive.

33. A(n) _____ is a substance that decreases the activity of the
central nervous system.

34. A(n) _____ is a substance that increases the activity of the
central nervous system.

35. _____ is a highly addictive stimulant found in the leaves of the
tobacco plant.

. . . .
Read each question, and write your answer in the space provided.

36. Explain how myelin speeds up the conduction of nerve impulses along an axon.

_____

_____

_____

37. Distinguish between the somatic nervous system and the autonomic nervous
system.

_____

_____

_____

_____

38. List the four basic types of chemicals that taste buds can detect.

_____

_____

39. Describe the harmful effects of smoke inhalation.

_____

_____

_____

Nervous System

In the space provided, write the letter of the description that best matches the term or phrase.

_____ **1.** neuron

_____ **2.** dendrite

_____ **3.** axon

_____ **4.** nerve

_____ **5.** membrane potential

_____ **6.** resting potential

_____ **7.** action potential

_____ **8.** synapse

_____ **9.** neurotransmitter

a. the difference in electrical charge across a cell membrane

b. part of a neuron that conducts nerve impulses

c. the membrane potential of a neuron at rest

d. nerve cell; transmits information throughout the body

e. bundle of neurons

f. part of a neuron that receives information from other neurons

g. a junction at which a neuron meets another cell

h. a signal molecule that transmits nerve impulses across synapses

i. nerve impulse

Write the correct term from the list below in the space next to its definition.

brain   hypothalamus   reflex
brain stem   interneurons   sensory neuron
central nervous system   motor neuron   spinal cord
cerebellum   peripheral nervous system thalamus
cerebrum

_____ **10.** carries motor responses from the central nervous system to muscles, glands, and other organs

_____ **11.** site of capacity for learning, memory, perception, and intellectual function

_____ **12.** consists of the brain and spinal cord

_____ **13.** relays sensory information

_____ **14.** dense cable of nervous tissue that runs through the vertebral column

_____ **15.** contains neurons that branch throughout the body

_____ 16. carries information from sense organs to the central nervous system

_____ 17. the body's main processing center

_____ 18. regulates breathing, heart rate, and endocrine functions

_____ 19. link neurons to each other

_____ 20. collection of structures leading down to the spinal cord

_____ 21. regulates balance, posture, and movement

_____ 22. a sudden, rapid, and involuntary self-protective motor response

In the space provided, write the letter of the description that best matches the term or phrase.

_____ 23. sensory receptor

_____ 24. retina

_____ 25. rod

_____ 26. cone

_____ 27. optic nerve

_____ 28. cochlea

_____ 29. semicircular canal

a. the lining of photoreceptors and neurons in the eye

b. aids in hearing

c. type of photoreceptor that responds best to dim light

d. runs from the back of each eye to the brain

e. helps maintain equilibrium

f. a specialized neuron that detects sensory stimuli

g. type of photoreceptor that enables color vision

Complete each statement by writing the correct term or phrase in the space provided.

30. The need for increasing amounts of a drug to achieve the desired sensation

is called _____ .

31. A drug that generally decreases the activity of the central nervous system is

called a(n) _____ .

32. A drug that generally increases the activity of the central nervous system is

called a(n) _____ .

33. Drugs that alter the functioning of the central nervous system are known as

_____ _____ .

34. _____ is a set of emotional and physical symptoms caused by removing a drug from the body of a drug addict.

35. _____ is a physiological response caused by repeated use of a drug that alters the normal functioning of neurons and synapses.

Hormones and the Endocrine System

In the space provided, write the letter of the term or phrase that best completes each statement or best answers each question.

_____ 1. Which of the following is NOT a characteristic of the endocrine system?
   a. Its chemical messengers are neurotransmitters.
   b. It coordinates all of the body's sources of hormones.
   c. Endocrine cells can release hormones directly into the bloodstream.
   d. Its chemical messengers bind to receptors.

_____ 2. Which of the following is a function of hormones?
   a. regulating growth
   b. maintaining homeostasis
   c. reacting to stimuli
   d. all of the above

_____ 3. An organ that secretes hormones directly into either the bloodstream or the fluid around the cells is called an
   a. exocrine gland.
   b. endorphin.
   c. endocrine gland.
   d. None of the above

_____ 4. The part of the brain that issues instructions to the pituitary gland is the
   a. brain stem.
   b. cerebellum.
   c. cerebrum.
   d. hypothalamus.

_____ 5. A hormone acts only on its target cell by
   a. stimulating a nerve cell.
   b. recognizing a receptor protein on or in the target cell.
   c. binding to a nerve cell.
   d. activating an enzyme in the blood.

_____ 6. Steroid and thyroid hormones form hormone-receptor complexes
   a. inside the cell.
   b. that bind to DNA.
   c. that activate or inhibit protein synthesis.
   d. All of the above

_____ 7. Amino-acid-based hormones
   a. are not fat soluble.
   b. can pass through cell membranes.
   c. change the shape of second messenger molecules.
   d. Both (b) and (c)

_____ **8.** Which gland acts as an emergency warning system in times of stress?
    a. adrenal gland   **c.** kidney
    b. ovary   **d.** thyroid gland

_____ **9.** Insulin and glucagon are two hormones that
    a. regulate the reproductive system.
    b. release calcium from bone.
    c. regulate the level of glucose in the blood.
    d. are secreted by the hypothalamus.

_____ **10.** If iodide salts are lacking in the diet, the thyroid gland
    a. becomes totally inactive.
    b. uses calcium to produce thyroid hormones.
    c. shrinks from lack of thyroid hormones.
    d. becomes greatly enlarged as it attempts to make more thyroid hormones.

_____ **11.** In a negative feedback mechanism, high levels of a hormone
    a. inhibit production of more hormone.
    b. stimulate production of more hormone.
    c. increase nerve impulses.
    d. decrease nerve impulses.

• • • • • • • • • • • • • • •
Circle T *if the statement is true or* F *if it is false.*

T   F   **12.** The effects of neurotransmitters last far longer than the effects of hormones.

T   F   **13.** Parathyroid hormone stimulates bone cells to release calcium into the bloodstream.

T   F   **14.** Some prostaglandins cause blood vessels to constrict, while others cause blood vessels to dilate.

T   F   **15.** Adrenal cortex hormones provide a faster, shorter-term response to stress than hormones from the adrenal medulla.

T   F   **16.** Insulin promotes the accumulation of glycogen in the liver.

T   F   **17.** Glucagon is an amino-acid-based hormone that is produced in the parathyroid gland.

T   F   **18.** Steroid hormones cannot pass through the cell membrane because they are not fat soluble.

T   F   **19.** Amino-acid-based hormones function without ever entering their target cells.

T   F   **20.** The hypothalamus and the anterior pituitary gland serve as a major control center for the endocrine system.

T   F   **21.** Epinephrine and norepinephrine are released in response to falling levels of calcium in the blood.

T   F   **22.** Parathyroid hormone is produced in the adrenal cortex.

T   F   **23.** Estrogens stimulate the development of secondary female sex characteristics.

T   F   **24.** The hypothalamus sends hormones to the anterior pituitary through a network of blood vessels between the two glands.

T   F   **25.** The islets of Langerhans produce insulin and glucagon.

T   F   **26.** Aldosterone helps retrieve potassium ions in the urine.

Complete each statement by writing the correct term or phrase in the space provided.

27. _____ _____ produce hormones.

28. _____ hormones are fat soluble.

29. _____ are a group of neuropeptides that are thought to regulate emotions and influence pain.

30. The release of _____ from the posterior pituitary stimulates uterine contractions and milk secretion.

31. When a steroid hormone binds to a receptor protein in a target cell's cytoplasm,

   a(n) _____ _____ complex is produced.

32. _____ _____ occurs when high levels of a hormone stimulate the output of more hormone.

33. High levels of the adrenal cortex hormone, called _____ , suppress the immune system.

34. The pituitary gland is made of two parts—the _____ and the

   _____ pituitary.

35. _____ are now rare in the United States because of the addition of iodine to salt.

36. _____ hormones regulate the body's metabolic rate and promote normal growth of the brain, bones, and muscles.

37. The pineal gland secretes the hormone _____ , and is thought to be involved in establishing daily biorythms.

Read each question, and write your answer in the space provided.

38. Describe the role of second messengers in relaying a hormone's message.

   _____

   _____

   _____

   _____

39. What two basic types of hormones does the hypothalamus send to the anterior pituitary, and what are their functions?

   _____

   _____

   _____

40. What is diabetes mellitus? If left untreated, how does diabetes mellitus affect the body?

_____

_____

_____

_____

_____

Questions 41 and 42 refer to the figure below, which shows a feedback mechanism. The number of hormone molecules represents the relative blood concentrations of Hormone A *and* Hormone B.

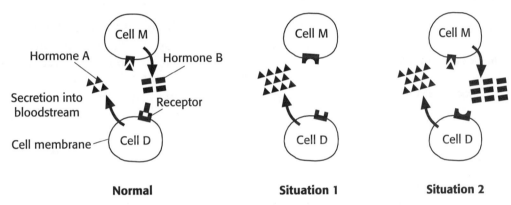

41. Which cell is defective in Situation 1? What happens to the hormone concentrations as a result of this defect?

_____

_____

_____

_____

42. Which cell is defective in Situation 2? What happens to the hormone concentrations as a result of this defect?

_____

_____

_____

_____

_____

Hormones and the Endocrine System

In the space provided, write the letter of the description that best matches the term or phrase.

_____ 1. hormone

_____ 2. endocrine glands

_____ 3. target cell

_____ 4. amino-acid-based hormones

_____ 5. steroid hormones

_____ 6. second messenger

_____ 7. negative feedback

_____ 8. hypothalamus

_____ 9. pituitary gland

_____ 10. adrenal glands

_____ 11. epinephrine

_____ 12. norepinephrine

_____ 13. insulin

_____ 14. glucagon

_____ 15. diabetes mellitus

a. structure of the brain that coordinates the activities of the nervous and endocrine systems

b. a hormone that lowers blood glucose levels

c. condition in which cells are unable to obtain glucose from the blood

d. a specific cell on which a hormone acts

e. a change in one direction stimulates the control mechanism to counteract further change in the same direction

f. substances that are secreted by cells and that act to regulate the activity of other cells

g. a molecule that passes a chemical message from the first messenger to the cell

h. a hormone that causes liver cells to release glucose

i. ductless organs that secrete hormones directly into either the bloodstream or the fluid around cells

j. one of two hormones released in time of stress; formerly called adrenaline

k. lipid hormones the body makes from cholesterol

l. secretes many hormones, including some that control endocrine glands elsewhere in the body

m. mostly water-soluble hormones

n. endocrine organs located above each kidney

o. one of two hormones released in time of stress; formerly called noradrenaline

Name_____ Date _____ Class _____

Reproduction and Development

In the space provided, write the letter of the term or phrase that best completes each statement or best answers each question.

_____ 1. Enzymes at the tip of the head of a sperm cell help the cell
   a. swim to an egg.
   b. penetrate an egg.
   c. find an egg.
   d. obtain energy for movement.

_____ 2. Sperm cells are produced by meiosis in the
   a. epididymis.
   b. vas deferens.
   c. seminiferous tubules.
   d. scrotum.

_____ 3. Sperm mature and become mobile in the
   a. epididymis.
   b. vas deferens.
   c. seminal vesicles.
   d. prostate gland.

_____ 4. Sperm are deposited in the female reproductive system by the
   a. vas deferens.
   b. seminal vesicles.
   c. scrotum.
   d. penis.

_____ 5. The gamete-producing organs of the female reproductive system are the
   a. corpus luteum.
   b. fallopian tubes.
   c. ovaries.
   d. seminiferous tubules.

_____ 6. After a female reaches puberty, one immature egg cell completes its development about every
   a. day.
   b. week.
   c. month.
   d. trimester.

_____ 7. On average, both the ovarian and menstrual cycles last
   a. 14 days.   **c.** 35 days.
   b. 28 days.   **d.** 45 days.

_____ 8. When the zygote reaches the uterus, it is a hollow ball of cells called a(n)
   a. embryo.   **c.** egg.
   b. blastocyst.   **d.** fetus.

_____ 9. It is especially important that pregnant women abstain from
   a. drugs prescribed by a doctor.
   b. exercise.
   c. alcohol.
   d. vegetables such as cauliflower and broccoli.

_____ 10. From 8 weeks until birth, a developing human is called a(n)
   a. fetus.
   b. embryo.
   c. zygote.
   d. blastocyst.

_____ 11. Genital herpes is caused by
   a. bacteria.
   b. the human immunodeficiency virus.
   c. a fungus.
   d. herpes simplex virus.

_____ 12. A chancre is a common symptom of
   a. PID.   c. gonorrhea.
   b. syphilis.   d. chlamydia.

_____ 13. Viral STDs
   a. can be treated and cured successfully using antibiotics.
   b. cannot be treated and cured successfully using antibiotics.
   c. cannot cause death if untreated.
   d. cannot be passed to a fetus.

• • • • • • • • • • • • • • • • • •

Circle T _if the statement is true or_ F _if it is false._

T F 14. Energy needed to propel sperm through the female reproductive system is supplied by ATP produced in the midpiece of the sperm.

T F 15. Semen is a mixture of sperm and fluid secretions from the seminal vesicles, the prostate gland, and the bulbourethral glands.

T F 16. The temperature in the scrotum is about 3°C higher than the rest of the body.

T F 17. The two phases of the ovarian cycle are the follicular phase and the luteal phase.

T F 18. When an ovum is released, it enters a fallopian tube, which leads from the ovary to the uterus.

T F 19. The entrance to the uterus is called the vagina.

T F 20. Implantation occurs when the blastocyst burrows into the lining of the uterus.

T F 21. The corpus luteum secretes progesterone, which increases the production of FSH and LH.

T F 22. A follicle is a cluster of cells that surrounds an immature egg in an ovary.

T F 23. The amnion interacts with the uterus to form the placenta.

T F 24. The menstrual cycle is influenced by the changing levels of progesterone and estrogen during the ovarian cycle.

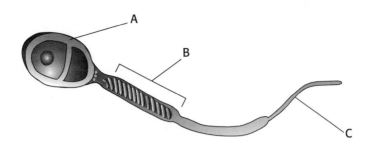

Complete each statement by writing the correct term or phrase in the space provided.

25. The structure labeled *A*, called the _____ , contains enzymes that help the sperm penetrate an ovum.

26. The energy that sperm need for movement is supplied by ATP produced in the

    _____ , labeled *B*. This energy powers the whiplike movements

    of the _____ , labeled *C*.

27. The testes are located outside the body cavity in an external skin sac called the

    _____ .

28. The forceful expulsion of semen through the penis is called

    _____ .

29. The _____ _____ begins when the anterior pituitary releases follicle-stimulating hormone and luteinizing hormone into the bloodstream.

30. When a follicle bursts, the mature egg cell is released in a process called

    _____ .

31. During _____ , the lining of the uterus is shed, blood vessels are broken, and bleeding results.

32. The embryonic membrane called the _____ encloses the embryo.

33. Most _____ STDs can be treated and cured using antibiotics.

34. By the end of the _____ trimester, a fetus is able to exist outside the mother's body.

35. _____ is a fatal disease caused by the human immunodeficiency virus (HIV).

36. _____ _____ _____ , which is one of the most common causes of infertility in women, is a condition linked to bacterial STDs.

Read each question, and write your answer in the space provided.

37. Trace the path that sperm travel once they leave the testes.

38. What happens during the luteal phase of the ovarian cycle?

39. What is fetal alcohol syndrome, and what are it symptoms?

40. Describe the events that occur early in the first trimester of pregnancy.

Question 41 refers to the figure at right, which shows the female reproductive system.

41. Identify the structures labeled *A–D*, and describe the functions of these structures.

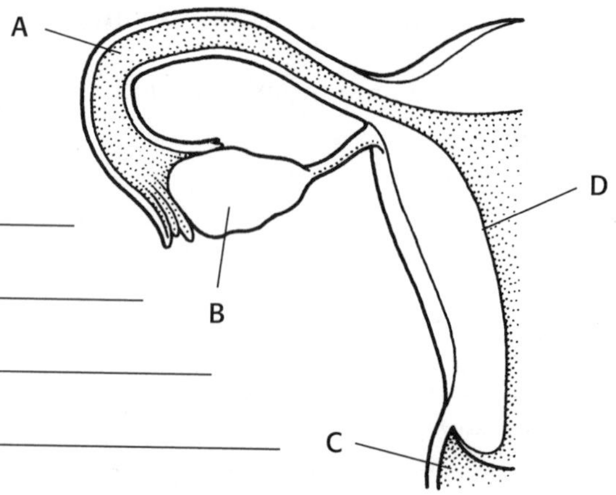

Name_____ Date _____ Class _____

Reproduction and Development

Write the correct term from the list below in the space next to its definition.

bulbourethral glands   prostate gland   seminiferous tubules

epididymis   semen   testes

penis   seminal vesicles   vas deferens

_____ **1.** produce a sugar-rich fluid that sperm use for energy

_____ **2.** long tube that connects the epididymis to the urethra

_____ **3.** the gamete-producing organs of the male reproductive system

_____ **4.** secretes an alkaline fluid that neutralizes the acids in the female reproductive system

_____ **5.** a mixture of exocrine secretions and sperm

_____ **6.** secrete an alkaline fluid that neutralizes traces of acidic urine in the urethra

_____ **7.** site where sperm are produced

_____ **8.** the male organ that deposits sperm in the female reproductive system during sexual intercourse

_____ **9.** site where sperm cells mature

Complete each statement by writing the correct term or phrase in the space provided.

10. The release of an ovum is called _____ .

11. The _____ are the gamete-producing organs of the female reproductive system.

12. The _____ _____ is the series of hormone-induced changes involved in the preparation and release of an ovum.

13. A mature egg cell is called a(n) _____ .

14. A(n) _____ _____ is a mass of follicular cells that secretes estrogen and progesterone.

15. A(n) _____ is a cluster of cells that surrounds an immature egg cell and provides the egg with nutrients.

16. The _____ is the hollow, muscular organ in which development occurs.

17. The shedding of the lining of the uterus is called _____ .

18. The _____ is the muscular tube that leads from the outside of a female's body to the uterus.

19. Each _____ _____ is a passageway through which an ovum moves from an ovary toward the uterus.

20. The series of hormone-induced changes that prepare the uterus for a possible

pregnancy each month is called the _____ _____ .

In the space provided, write the letter of the description that best matches the term or phrase.

_____ 21. gestation

_____ 22. placenta

_____ 23. cleavage

_____ 24. fetus

_____ 25. embryo

_____ 26. blastocyst

_____ 27. implantation

_____ 28. pregnancy

a. a hollow ball of zygotic cells

b. period of development; pregnancy

c. a series of internal divisions that the zygote undergoes in the first week after fertilization

d. a developing human during the first 8 weeks after fertilization

e. burrowing of the blastocyst into the lining of the uterus

f. a developing human from the eighth week of pregnancy until birth

g. a structure through which the mother nourishes the fetus

h. divided into three trimesters

Complete each statement by writing the correct term or phrase in the space provided.

29. _____ is a bacterial STD that causes painful urination and a discharge of pus from the penis in males. In females, it sometimes causes a vaginal discharge.

30. _____ is a serious bacterial STD that usually begins with a small, painless ulcer called a chancre 2–3 weeks after infection.

31. The symptoms of _____ , a bacterial STD, are similar to those of a mild case of gonorrhea and can cause scar tissue in infected fallopian tubes, leading to infertility.

32. One of the most common causes of infertility in women is _____

_____ _____ , which is a severe inflammation of the uterus, ovaries, fallopian tubes, or abdominal cavity that results from an untreated bacterial STD.

33. _____ _____ is a viral STD that includes periodic outbreaks of painful blisters in the genital region, flulike aches, and fever.